Lesso
Longford: the trial

CONTENTS

Preface **3**

Chapter 1: Introduction **5**

Chapter 2: The charges **11**

Chapter 3: Esso's defence **27**

Chapter 4: Sentencing **33**

Chapter 5: Learning from Longford **41**

Chapter 6: Corporate manslaughter **49**

Chapter 7: Conclusion **55**

Appendix: The sentence **59**

Index **69**

SUBSCRIPTIONS

The Journal of Occupational Health and Safety — Australia and New Zealand is published every two months by CCH Australia Limited, ABN 95 096 903 365. It is available by subscription.

Subscription rates (Aust and NZ)

$A583 per year (incl GST), $NZ592 per year (plus GST and shipping charges)

To register subscriptions:

Australia, Pacific, Africa, South America
CCH Australia Limited
GPO Box 4072
Sydney, NSW 2001

New Zealand
CCH New Zealand Limited
24 The Warehouse Way
Northcote 1309, Auckland 10

Singapore and Brunei
CCH Asia Pte Limited
#11-01 RCL Centre
11 Keppel Road
Singapore 089057

Japan
CCH Japan Limited
Towa Misakicho Bldg 5F
3-6-2 Misakicho
Chiyoda-ku, Tokyo 101-0061

United States
CCH Incorporated
4025 West Peterson Ave
Chicago, Ill. 60646

Canada
CCH Canadian Limited
90 Sheppard Avenue East
Suite 300
North York
Ontario MZN 6X1

United Kingdom, Europe
Croner. CCH Group Limited
145 London Road
Kingston upon Thames
Surrey KT2 6SR

Back issues

Back issues are available on application, subject to availability.

Individual issues

Individual issues are available for purchase on request, subject to availability.

Advertising

Advertising enquiries should be directed to the Advertising Manager ☎ (02) 9857 1502. Camera-ready artwork is required.

The advertisement of a product or service in this journal does not imply endorsement.

The Publisher reserves the right to accept or reject advertising material. Advertisers are required to satisfy the requirements of the *Trade Practices Act* and the legislation relating to advertising in each State.

Notes for contributors

Please refer to the back page.

IMPORTANT DISCLAIMER

ISSN 0815-6409

Reprinted.. March 2005

PREFACE

This issue of the journal is devoted entirely to my analysis of the Esso Longford criminal trial. The analysis is written as a book, but it is too short for publication as a conventional book. On the other hand, it is too long to be a single journal article. Hence, this somewhat innovative publication strategy.

My earlier book, *Lessons from Longford*, detailed the causes of the 1998 Esso gas plant explosion at Longford, near Melbourne, but that is not the end of the story, for the legal consequences of the explosion have continued to unfold.[1] Moreover, these legal consequences are just as worthy of attention as the explosion itself, since they are in many respects history-making. The present publication is concerned with the criminal trial which took place in 2001. A civil trial was conducted during 2002, with a decision due to be handed down at the end of the year or early in 2003. The decision may in fact coincide with the appearance of this work and it is therefore important to distinguish the two trials.

The criminal trial, under the *Occupational Health and Safety Act 1985*, found that Esso had failed in its duty of care to its employees. In the civil action, the allegation is that Esso had a duty of care to its ultimate customers, and that its failure to ensure uninterrupted supply amounted to a breach of this duty. If the civil action succeeds it will establish a precedent with far-reaching commercial consequences. Companies which provide essential services will need to insure against the possibility that they may be unable to maintain supply and these insurance costs will necessarily be passed on to customers.[2] Whatever the outcome of the civil trial, the matter may well be appealed to the High Court, and it could be years before the final decision is known.

To reiterate, this book is about the criminal trial and its significance. My thanks to Richard Johnstone for his careful reading of the manuscript and for stimulating discussion of the issues.

Andrew Hopkins
Australian National University, Canberra
(Andrew.Hopkins@anu.edu.au)
December 2002

1. Hopkins, A. *Lessons from Longford: the Esso gas plant explosion*. Sydney: CCH Australia Limited, 2000. Available from CCH Customer Support on ☎1300 300 224 or email support@cch.com.au.
2. Bartholomeusz, S. Can we risk the problems in insuring against breakdown of supply? *The Age*, Melbourne, 9 October 2002, p 3.

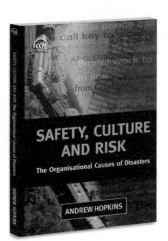

CHAPTER 1

Introduction

"What [happened] ... was no mere accident. To use the term 'accident' denotes a lack of understanding of responsibility and a lack of understanding of cause."[1] So said the Judge at the conclusion of Esso's trial. An explosion in 1998 at the company's gas plant at Longford, near Melbourne, had killed two men, injured eight others, and cut gas supplies to the city of Melbourne for two weeks. Esso claimed at the trial that what happened was an unavoidable accident. The Judge rejected this; the cause of the incident, he said, was "grievous, foreseeable and avoidable".[2]

The Longford explosion has aroused intense interest, both in Australia and internationally. The third edition of a classic work on major accidents published in the United Kingdom devotes a chapter to Longford.[3] And following the prosecution, a long article on the subject appeared in the *New York Times*. "Try as it might [the article began], Exxon Mobil cannot seem to keep a 1998 gasworks accident — and the way it responded to it — from coming back to bite it again and again."[4]

Why such interest? In part because the incident came as such a shock. According to one European observer: "What must also give us pause is that the accident happened in a company which has a very good reputation in some other parts of the world. It shows yet again how fragile such reputations are."[5] Kletz also muses about this:

> "Exxon, the owner of the Esso plant, had an enviable and well-deserved reputation for its commitment to safety. Was Longford a small plant in a distant country that fell below the company's usual standards or did it indicate a fall in standards in the company as a whole? Perhaps a bit of both."[6]

Not surprisingly, the oil and gas industry around the world has drawn what lessons it can from the incident, but other industries using rather different technologies are also taking note. The nuclear industry in the UK, for example, is aware that organisational change needs to be subjected to the same systematic study which is now given to physical changes to plant and processes. Longford has provided it with a powerful example of what can go wrong when organisational change is not well managed.[7]

1. The sentence, para 7 (see Appendix at p 60).
2. The sentence, para 10 (see Appendix at p 61).
3. Kletz, T. *Learning from accidents* (3rd ed). Oxford: Gulf, 2001.
4. *New York Times*, 6 December 2001.
5. Hale, A. Review of "Lessons from Longford". *Safety Science* 2001, 38: 258.
6. Kletz, T. Review of "Lessons from Longford". *Chemical Engineering Progress* 2001, 97(9): 78-79.
7. Kletz, T. Personal communication.

In Australia, too, there is continuing interest in the Longford explosion in many industries, both high-tech and otherwise. While some companies are quick to distance themselves from Esso, believing that they manage safety considerably better, many others have the uneasy feeling that "there but for the grace of God go we". More than one director has said to me: "We believe that we manage safety well, but so did Esso. How can we be sure that we are not about to have a 'Longford' here?"

Shell Australia provides a stark illustration that Esso is not alone. Shell is fabled worldwide for its commitment to safety. However, an audit and investigation of the Shell refinery near Melbourne which was carried out by the Victorian OHS agency (the Victorian WorkCover Authority) — as it happens a few weeks after the Longford incident — revealed serious deficiencies in Shell's fire protection system. An earlier audit had also disclosed serious shortcomings in the system. Shell was therefore prosecuted, and the magistrate observed that, although the offences had not resulted in catastrophe, this did not diminish their seriousness. He fined the company a total of $225,000.[8] If both Esso and Shell can fail so dramatically, what company can feel safe?

My earlier book, *Lessons from Longford*, drew on the proceedings of the Longford Royal Commission to identify the organisational causes of the Esso explosion.[9] But since that time, events have continued to unfold. Esso has been fined $2m, the largest fine ever imposed in Australia for an OHS offence. Moreover, 10 victims of the disaster have been awarded more than $1m in compensation by the Victorian Supreme Court.[10] Overshadowing all of this from a financial point of view, a civil action for over $1b was commenced, stemming from the two-week loss of gas supply to Victorian customers.[11]

There has also been a profound regulatory response to the disaster. Companies in Victoria which have to manage major hazards are subject to stringent new legal requirements and an ultimately unsuccessful attempt was made by the Victorian Government to enact corporate manslaughter legislation.[12]

The purpose of the present work is to explore some of these developments and, in particular, the trial.[13] The Esso matter was heard in the Victorian Supreme Court, the highest court in the State, and was the first ever OHS trial at that level. The outcome was an enormous morale boost for WorkCover Victoria (the agency which brought the case), its Chief Executive announcing that "the verdict and sentence confirms that the Authority is formidable in its capacity to enforce the legislation it administers".[14] The proceedings, including the sentence and the comments made by the Judge in passing sentence, provide an important insight into how the law can be expected to deal with high profile cases of this nature in the future.

This study addresses a number of questions:

— Why did the Judge say so emphatically: "What happened was no mere accident"?

— How culpable was Esso?

— Why was Esso charged on so many counts — 11 in all?

8. WorkSafe Victoria. *Recent prosecutions, cases heard 1/1/2000 to 31/12/2000*. Melbourne: WorkSafe Victoria, 2001.

9. Hopkins, A. *Lessons from Longford: the Esso gas plant explosion*. Sydney: CCH Australia Limited, 2000. Available from CCH Customer Support on ☎1300 300 224 or email support@cch.com.au.

10. *The Age*, 20 December 2001. Eighteen other compensation claims were earlier settled out of court on a confidential basis. Following that settlement, the Judge announced that details of any further settlements would need to be made public. Ruling no 16, 3 October 2001.

11. By mid-2002 the figure had been reduced to $500m. *Canberra Times*, 1 June 2002.

12. Victorian Government. *Occupational Health and Safety (Major Hazard Facilities) Regulations 2000*.

13. Pre-trial proceedings occurred in 2000. The trial occupied the first half of 2001. Sentence was handed down on 30 July 2001 — [2001] VSC 263. Judge: Cummins J; for the prosecution: R Richter QC; for the defence: M Titshall QC.

14. Mountford, W. *WorkWords*. Publication no 34. Melbourne: Victorian WorkCover Authority, September 2001, p 2.

- Are OHS offences crimes?
- How effective was Esso's defence strategy?
- Why did Esso *not* blame its control room operators at the trial, when it had blamed them at the Royal Commission?
- What does "practicably preventable" mean?
- What meaning can be given to the concept of corporate remorse?
- What kinds of prior corporate convictions count for the purposes of sentencing?
- Did the trial produce any new findings about the causes of the disaster?
- How did the confusion between "risk" and "hazard" affect the trial?
- To what extent has Esso learnt the lessons of Longford?
- Would Esso have been at risk of conviction under the corporate manslaughter legislation which was proposed for Victoria?
- Why was the trial held in the Supreme Court?

This short work is written for a readership without particular knowledge of the ways of courts. Court proceedings are long-winded and dreary. This trial went for several months and the press rapidly gave up reporting it. There is an obvious need, therefore, for a lay account of the trial and its implications. The details of how a large corporation was charged with criminal offences and how it defended itself are really quite fascinating. Lawyers may find the discussion elementary, but I hope they will find no inaccuracies. Safety management professionals will find some of the conclusions thought-provoking. Company executives will learn something about what the law now expects of them and the potential legal consequences of failure to maintain a safe workplace.

The analysis is intended to be self-contained in the sense that it does not presume a reading of my earlier work, *Lessons from Longford*. However, I have not wanted to repeat the discussion in that book here and readers will find it useful to refer to it to fill in various gaps.

The latter part of this introduction deals with the differences between a Royal Commission and a criminal trial. Chapter 2 deals with the charges, Chapter 3 with Esso's defence, and Chapter 4 with the sentence. The complete text of the sentence is contained in an Appendix. Following conviction, Esso put its safety achievements on display in order to persuade the Judge to impose a more lenient sentence. Chapter 5 summarises what can be learnt from this about safety improvements at Longford since the explosion. It discusses the Judge's response to Esso's claims about its safety improvements and draws conclusions about the extent to which Esso has learnt the lessons of Longford. It notes briefly how WorkCover Victoria is applying some of those lessons. Chapter 6 discusses the defeated Victorian corporate manslaughter Bill and speculates about whether Esso might have been found guilty under it, had it been law at the time of the explosion. Chapter 7 sets out some answers to the questions listed above.

Comparing the trial and the Royal Commission

The first official response to the Longford incident was a Royal Commission of inquiry; the criminal trial came later. It is worth outlining the differences between these two types of legal proceedings before embarking on the details of the trial.

The basic purpose of the Royal Commission was to get to the bottom of what happened. The Commission was guided by a set of questions, provided in its terms of reference, which aimed to identify the causes of the explosion and fire. Its function was to answer these questions to the best of its ability. It was headed by a former High Court Judge and a highly experienced chemical engineer, and it had a staff of expert investigators. Its report is an authoritative and wide-ranging account of what happened and why.[15]

15. Dawson, D and Brooks, B. *The Esso Longford gas plant accident: report of the Longford Royal Commission*. Melbourne: Government Printer for the State of Victoria, June 1999.

The Commission was not a forum in which any party could be held accountable or punished. However, it was specifically asked to consider whether any breach of legislation by Esso had contributed to the explosion and it duly found that Esso had violated the Victorian *Occupational Health and Safety Act 1985*. Given that the procedures which are followed in an inquiry with respect to standards of evidence and the rights of defendants are not as stringent as those which are followed in a criminal trial, this finding did not amount to a criminal conviction. Nor could it be used in the subsequent trial.[16]

The Esso trial was a very different affair. Trials are not designed to provide an account of what happened; they aim to determine the guilt or otherwise of the defendant (in this case a company, Esso Australia) and to determine appropriate punishment. This means that complex factual issues which are not directly relevant to the charges may remain unresolved. This point will be elaborated in what follows.

A trial is a contest. Counsel for the prosecution jousts with counsel for the defence according to an intricate set of evidentiary and procedural rules, and the major function of the judge during the trial is to ensure that these rules are scrupulously followed. Moreover, both the prosecution and the defence counsel are seeking to persuade the jury of the merit of their case. Both owe important duties to the court, for example, not to mislead it, but neither is expected to present a balanced view: the prosecution can be expected to present the defendant in the worst possible light, while defence counsel presents the best possible interpretation of the defendant's behaviour.

Given these roles, prosecution and defence counsel can be expected to dramatise their case in various ways, to resort to oratorical tricks and to play on the emotions and prejudices of jurors in an attempt to win them over. Additionally, counsel may accuse each other of such things in order to try to undermine the impact of the opposing case. For example, counsel for the prosecution told the jury in his opening address that workers at Longford had been unknowingly "sitting on a bomb" — not language which could be expected to appear in the report of a Royal Commission, no matter how appropriate it may have been. And again, counsel for Esso suggested that the prosecution had described Esso as a large company with United States connections, thereby setting it up as a "tall poppy". He urged the jury not to be influenced by the stupid and negative Australian tendency to cut down tall poppies and to give Esso a "fair go".[17]

It is not the role of the judge to declare the winner; that is the jury's function. More precisely, the jury is called on to declare whether, in its view, the defendant is guilty beyond reasonable doubt. It does not have to give reasons for its decision. The jury in the Esso trial was therefore not required to pass comment on the many arguments or theories put to it or to indicate which it found most convincing.

In view of these features, a boxing match may be a more applicable metaphor than a jousting contest. The two counsel are like boxers seeking to outwit and outperform each other, the trial judge is the referee, ensuring fair play, and the jury is the panel which scores the fight and determines the outcome, but is not required to give reasons for its decisions.

The point I wish to draw from this discussion is that none of the participants in the Esso trial had the responsibility of providing a detailed and balanced account of what happened. The trial therefore did not give rise to a set of evidence-based findings about the causes of the disaster in the way that the Royal Commission did. In principle, a trial can provide no such finding, beyond the guilt or innocence of the accused.

16. Nor could admissions made by Esso in its final submission to the Royal Commission be treated as admissions for the purposes of the trial. Ruling no 8, 20 March 2001.

17. *The Age*, 9 March 2001.

There is, however, one important way in which the Esso trial went beyond the Royal Commission. Unlike a Royal Commission, a trial is a forum in which culpability is assessed. If a jury decides that a defendant is guilty, the second major function of a judge comes into play — to determine an appropriate sentence. At this stage a judge may make some considered comments based on the circumstances of the case. These are not findings about what happened but comments about how blameworthy the behaviour was. These comments, together with the magnitude of the sentence which the judge imposes, provide an authoritative statement of blameworthiness. Of course, blameworthiness cannot be evaluated in an objective way. But neither can the allocation of blame be arbitrary or idiosyncratic. Courts must ultimately appeal to wider community standards to legitimate their sentencing decisions. Sentencing therefore provides an indication both of how courts view the behaviour in question and, at least to some extent, of wider community views.

The upshot of all of this is that the analysis of the trial which follows cannot be expected to advance our understanding of what happened at Longford beyond the findings of the Royal Commission. It can, however, contribute to an understanding of how and why the Court reacted as it did and what legal and community expectations are in relation to the behaviour of large corporations.

CHAPTER 2

The charges

Esso faced a bewildering array of charges — 11 in all.[1] A number of charges were directed to different steps in the hazard management process, that is, hazard identification, risk assessment and risk control, and several charges concerned training failures. There was arguably some duplication in the charges and one charge was rendered meaningless as a result of the confusion between hazard and risk. The main purpose of this chapter is to describe these charges and to explore their significance. The reasons why Esso faced so many charges will also be addressed.

All of the charges alleged that Esso failed to do what was practicable to prevent the explosion. The meaning of practicability is therefore a crucial issue which is discussed towards the end of the chapter.

The chapter begins, however, by explaining how the case came to be heard in the State's highest court.

Why a Supreme Court trial?

The charges were initially brought before a magistrate to determine whether there was a case to be heard. Esso chose not to contest the case at this point and the company was thereupon committed for trial in the County Court. (The County Court is one level above the Magistrates' Court in the hierarchy of courts.) In the normal course of events the County Court is where the trial would have taken place. However, because of the exceptional nature of the case, the prosecution applied to have the matter heard at the highest level — in the State's Supreme Court. Esso opposed this application but the Supreme Court acceded to it on the grounds that the matter was of substantial interest to both the parties and the public and that it needed to be heard in an authoritative court. The Judge's ruling on the application was as follows:

> "The matter is of substantial significance to the parties involved and to the persons affected by it. It is a matter of legal significance, and the terms of the Act are terms which may call for authoritative, that is to say, legally binding construction. It seems to me that it is appropriate that an authoritative court, that is to say, a legally binding court, should be seized of those matters. Further, the matter is of substantial public interest, and by that I do not mean the quantum of publicity but, rather, the inherent nature of the matters are matters of what the court and the law regards as of substantial public interest."[2]

1. It was initially charged on 45 counts — see discussion later.
2. Ruling no 1, 7 August 2000.

This was the first ever trial under the Victorian *Occupational Health and Safety Act 1985* (the OHS Act) to be heard in the Victorian Supreme Court. The importance of this was that it would set precedents for lower court trials about how corporate misconduct of this nature should be dealt with and, in particular, the level of fines which were appropriate.

Where did the prosecution focus in the incident sequence?

The investigation by the Royal Commission revealed that the explosion was the culmination of a long chain of events which began nearly 24 hours earlier. For many hours, plant operators ran the system outside its design envelope, that is, at a combination of temperatures, pressures and flow rates for which it was not designed. To do this they had to ignore a variety of alarms in the control room which were intended to alert them to the need for action to bring the system back within its designed operating range. Operation in this alarm mode eventually led to the automatic shutdown of certain pumps which circulated warm "lean" oil throughout the plant. This occurred at 8.20 am on Friday, 25 September 1998 — nearly four hours before the explosion.

The shutdown of the warm oil pumps meant that the cooling processes used at the plant were now unchecked and the temperature of certain large steel tanks — heat exchangers — fell well below safe limits. The tanks became brittle with cold. The operators were unable to get the pumps restarted for some hours but finally succeeded, and when the warm oil reached the steel tanks again one of them shattered, releasing a vast amount of gas which exploded when it reached a nearby flame source.[3]

The Royal Commission devoted attention to every part of this extended incident sequence. However, the prosecution took the view that it did not matter why the pumps shut down and it focused exclusively on events after the shutdown, that is, the dangerous embrittlement of the heat exchangers and the disastrous reintroduction of warm oil. Its entire case against Esso was built around the company's failure to envisage the consequences of the pumps shutting down or to plan adequately for this possibility.[4]

There were at least three reasons for the prosecution to limit itself in this way. First, the earlier part of the sequence was extremely complex and the precise mechanisms remained partially obscure, even to the Royal Commission. To have attempted to lead the jury through this material may well have generated such confusion in their minds as to require them to give Esso the benefit of the doubt.

Second, as Kletz pointed out:

> "All pumps are liable to stop from time to time. There are many possible causes such as mechanical failure, loss of power supply, low level in the suction vessel and high level in the delivery vessel ... The precise reason why the flow stopped on 25 September 1998 is of secondary importance. Next time the pump will stop for a different reason ... Whatever the reason for the pump stopping, the consequences of the interruption of flow were of major importance."[5]

Third, the consequences of pump stoppage were far more direct and foreseeable than the consequences of allowing the plant to operate in alarm mode. Esso's failure to envisage the consequences of the interruption of warm oil flow was therefore far more culpable. Since the aim of the prosecution was to demonstrate culpability, it made sense to limit itself in this way.

3. These were the findings of the Royal Commission.
4. Counsel told the jury at one stage: "It is vital to remember that the prosecution case is that it does not matter why the heating medium was lost." This quotation is from the transcript. Page references for material drawn from the transcript will not be provided.
5. Kletz, T. *Learning from accidents* (3rd ed). Oxford: Gulf, 2001, p 268.

What were the charges?

Section 21(1) of the OHS Act imposes a duty on employers to "provide and maintain so far as is practicable for employees a working environment that is safe and without risks to health". This is what the Royal Commission had said that Esso had failed to do and this is what the prosecution set out to prove.

A defendant in a criminal trial is entitled to know precisely what is being alleged. The allegation that an employer has failed to provide a safe workplace is obviously far from precise. The prosecution must therefore specify precisely how the defendant has failed. If there is more than one way, then each failure must be the subject of a separate charge or count.[6] Thus it was that Esso was charged on a number of counts, each of which took the following form:

"Count 1: Esso Australia Pty Ltd at Longford in the said State [Victoria] between the 1st day of January 1993 and the 25th day of September 1998 being an employer failed to
provide and maintain so far as was practicable for employees a working environment that was safe and without risks to health
by failing to
make arrangements for ensuring so far as was practicable safety and absence of risks to health in connection with the use and handling of plant and substances
in that it
failed to conduct any adequate hazard identification at Gas Plant One."
(italics and bold emphasis added)

This is a complex formulation which requires explanation. The first italicised section of the count is the general duty. Every count on which Esso was charged began with these words. Section 21(2) of the OHS Act lists a number of ways in which employers may contravene their general duty, and in order to give greater specificity to each count the prosecution added the second italicised set of words, drawn from this list (in this case, concerning "the use and handling of plant and substances"). But even these words are too vague to give a defendant much indication of what is being alleged. Hence, the final set of words, known as particulars, were added (reproduced here in bold type). These at last make clear just what it is that the prosecution alleged Esso should have done but didn't — in this case, carry out hazard identification.

In the end there were 11 such counts. In summary form they were:

1. Inadequate hazard identification.
2. Failure to conduct an adequate risk assessment.
3. Failure to maintain plant in a safe condition.
4. Inadequate procedures to deal with loss of warm oil circulation.
5. Failure to adequately train employees to respond to a loss of warm oil.
6. Failure to provide the means for ensuring that items of plant operated at a safe temperature.
7. Failure to adequately train employees regarding the risks associated with abnormally cold plant.
8. Failure to adequately train supervisors regarding the risks associated with abnormally cold plant.
9. Failure to monitor conditions.
10. Failure to ensure that the heat exchanger was protected from thermal shock.
11. Failure to ensure the safety of persons other than employees.

These counts are elaborated on below.

1. Inadequate hazard identification

First, and fundamentally, the company was charged with inadequate hazard identification. There are two conceptually distinct aspects to the hazard identification process. The first is that underlying hazards must be identified: the fact that metal which is chilled to a low enough temperature can become

6. The issue of multiple counts is discussed more fully later.

brittle with cold and hence susceptible to fracture was an underlying hazard in a plant such as Gas Plant 1. Second, ways in which underlying hazards might give rise to major incidents must be discovered, for example, the possibility that a heat exchanger could become brittle with cold and fracture, releasing a cloud of flammable material.[7] Any systematic hazard identification process will naturally identify both the underlying hazards and the incidents to which they might give rise.

The standard hazard identification process in the petrochemical industry is the HAZOP (hazard and operability study). In such a study, a team of experts carries out a "what if?" analysis using a piping and instrumentation diagram for the plant. At each junction or decision point of the diagram the team asks a series of questions, for example, what if the flow at this point was too high, too low, the temperature too high, too low, the pressure too high, too low, etc? The team examines the consequences of such deviations from normality and the kinds of incidents which might arise. In this way, hazards lurking in the plant can be identified and action taken to ensure that they do not give rise to an incident.

Gas processing equipment today is often designed to withstand temperatures that are well outside the normal operating range. But plant built before the early 1970s is less likely to have been designed in this failsafe way and depends to a greater extent on operator interventions and automatic shutoff mechanisms for its safety. Esso's parent, Exxon, was well aware of the dangers of embrittlement when metal vessels fall below their design temperatures and was particularly concerned about the vulnerability of its older gas plants to this problem. In June 1994, it issued HAZOP guidelines to all its affiliates, including Esso Australia, about the need to be especially alert to the problem of embrittlement. Exxon recommended to its affiliates that they carry out HAZOPs retrospectively on all plant which was more than 20 years old. There were three gas plants at the Longford site and Gas Plant 1, the plant at which the explosion occurred, was built in 1969 and should therefore have been HAZOPed. It wasn't, and Esso did not explain this omission either at the trial or the Royal Commission.

The Exxon HAZOP guidelines included the following comments:

> "[These guidelines] should be useful if steel equipment is exposed to low temperature operation. Low temperature is typically defined as temperature below 0°C so nearly all equipment is of concern when started up or operated at low ambient temperatures. Also any equipment handling materials which could auto-refrigerate to 0°C or below at atmospheric pressure should be reviewed for cold safety concerns. [NB: Gas Plant 1 could auto-refrigerate.] Caution. Equipment has been known to fail from brittle fracture above 0°C. If the type of steel toughness cannot be verified and/or the minimum design temperature is not known, then detailed review for fitness for service above 0°C may be needed."

The document lists various prompt questions, including: "Can the equipment get colder than the specified critical exposure temperature … due to … loss or reduction in heat …?"

This, of course, is what happened at Gas Plant 1. The guidelines in fact contained a fairly detailed description of the precise scenario which occurred at Longford on 25 September 1998, leading counsel for the prosecution to describe the Exxon document as "the script for the catastrophe".

But suppose Esso had carried out the HAZOP. Is there any guarantee that it would have discovered the problem? The prosecution produced convincing evidence at the trial that it would have. One of Esso's witnesses was asked to carry out an imaginary HAZOP in the witness box. He was forced to admit that the problem of embrittlement was "blindingly obvious" and that any competent HAZOP team would "see it coming at them like a steam train".

7. Victorian WorkCover Authority Major Hazards Division. *Hazard identification under the Occupational Health and Safety (Major Hazard Facilities) Regulations* (MHD GN-13) provides a useful discussion of this distinction. Website at www.workcover.vic.gov.au/vwa/home.nsf/pages/so_majhaz_guidance.

Hazard identification is part of the now widely accepted risk management approach. Many Regulations made under OHS statutes adopt this approach and make hazard identification the starting point for ensuring the health and safety of employees. In particular, hazard identification is the starting point for the Victorian *Occupational Health and Safety (Major Hazard Facilities) Regulations 2000* (MH Regulations). It is not explicitly required under the OHS Act, or under other Australian OHS Acts of its vintage. It is, however, an implicit requirement, and the conviction reinforces the fact that employers must adopt a structured system for identifying hazards, especially in hazardous environments, in order to comply with their general duty of care.

2. Failure to conduct an adequate risk assessment

The second charge brought against Esso concerned risk assessment. There was considerable confusion concerning this charge, stemming, ultimately, from confusion in the safety literature about the meaning of "risk". In order to understand the Court's confusion, we first need to explore the meaning of risk management.

The risk management approach is embodied in many OHS Regulations, including the Victorian MH Regulations. It involves three basic steps: hazard identification; risk assessment; and risk control.

The first of these, hazard identification, has already been discussed. Once a hazard and its associated incident have been identified, the next step is to analyse the risk of such an incident. Risk is understood in this context to refer to both likelihood and severity: the higher the likelihood of an incident, the greater the risk, *and* the more severe the consequences of the incident, the greater the risk. Risk assessment thus involves an evaluation of both the likelihood and the consequences of any incident which has been thrown up in the hazard identification process.

The final step in the risk management process is to set in place controls for dealing with the hazard. Just how elaborate these controls need to be depends on the outcome of the risk assessment process, that is, on how likely the incident and how severe its consequences.

I shall call this the "OHS formulation" of risk management because it is the formulation which has developed in the OHS literature. The point to be stressed about this formulation is that it makes a careful distinction between hazard and risk.

Unfortunately the word "risk" is not always used in the manner described above. In some contexts it means "hazard", *as well as* referring to the assessment of likelihood and severity.

The problem is well illustrated in the Australian standard on risk management (AS/NZS 4360), which has been developed to cover a broader range of risks, such as financial, legal, political and commercial risks, as well as risks to the health and safety of employees.[8] The three relevant steps described in this standard are risk identification, risk analysis and risk treatment.[9] The two formulations are set out below to facilitate comparison:

OHS formulation	*AS/NZS 4360*
Hazard identification	Risk identification
Risk assessment	Risk analysis
Risk control	Risk treatment

Although the terms used in the Australian standard are not quite the same as those used in the OHS formulation, the definitions provided in the standard reveal that they refer to the same steps.

Most importantly for present purposes, the first step, risk identification, is defined in the standard as "the process of determining what can happen, why and how". This is exactly the meaning of hazard identification in the OHS formulation. In short, when the word "risk" is used as part of the phrase "risk identification" in the standard, it means "hazard".

8. AS/NZS 4360-1999: *Risk management*. Sydney: Standards Australia, 1999.

9. The standard includes three other steps in the process, but these are not relevant to the present discussion.

On the other hand, the standard defines risk as "the chance of something happening that will have an impact upon objectives. It is measured in terms of consequences and likelihood". Furthermore "risk analysis" is defined as determining "how often specified events may occur and the magnitude of their likely consequences". Risk here is being used in the way it is used in the OHS formulation. The standard, therefore, uses "risk" to refer to two different things. This is the essence of the problem.

Consider now the "periodic risk assessments" which Esso undertook. What did "risk" mean in this context? Did it refer to severity and likelihood of consequences — or did it mean hazard?

Periodic risk assessments were carried out every three years in some circumstances and every five in others. The assessment team first developed a list of things which might go wrong at the facility, based on their own experience and on incident data from other similar plants. These were termed "scenarios". The team then worked through each of these scenarios asking "what if?" questions — what would happen at this plant if this scenario occurred — and identified as systematically as they could all the possible consequences.[10] For example, a previous periodic risk assessment done at Longford worked through one cold temperature scenario. The team imagined a situation in which a valve nozzle made of carbon steel, rather than low temperature steel, was used by mistake in a piece of low temperature equipment. They concluded that it could become brittle and therefore dangerous under certain circumstances. Recommendations were made to avoid this possibility.

A periodic risk assessment was carried out at Gas Plant 1 in 1994. It was limited in scope, however, because of the plan at that stage to conduct a HAZOP of Gas Plant 1 in 1995 and it did not cover matters which the proposed HAZOP would cover.[11] (The proposed HAZOP was never in fact carried out.)

It is clear from the above description that the periodic risk assessment is in fact a hazard identification exercise in that it is concerned with underlying hazards and ways in which they could result in major incidents. It is not as systematic as a HAZOP and is no real substitute for a HAZOP, but it covers matters which might not be captured in a HAZOP and is clearly a valuable addition to the armoury of hazard identification procedures.

The fact that Esso had called this process a "risk assessment" led to considerable confusion at the trial. In charging Esso with failure to carry out an adequate risk assessment, the prosecution had in mind that Esso's 1994 periodic risk assessment had been limited in the way described above and had not considered the possibility of embrittlement of the heat exchanger. In effect, it was charging Esso with failure to carry out this alternative form of hazard identification adequately. But it is hard to see why the decision to limit the periodic risk assessment in this way was so culpable. In the circumstances at the time, namely that a HAZOP had been scheduled, this seems a perfectly reasonable decision.

The main problem, however, was that the prosecution asserted that the periodic risk assessment was not a hazard identification process:

> "It is a complete and absolute nonsense to say that a periodic risk assessment can stand in the place of a hazard identification process for the purpose of identifying hazards. They are two different beasts."

However, if the periodic risk assessment was not a form of hazard identification, what was it? Perhaps in the mind of the prosecution it was the step in the risk management process which follows hazard identification, that is, assessment of the likelihood and severity of the incident which has been identified as a possibility. That, in fact, is what the Judge understood the prosecution to be saying:

10. This description of a "periodic risk assessment" is taken from evidence given to the Royal Commission.

11. Dawson, D and Brooks, B. *The Esso Longford gas plant accident: report of the Longford Royal Commission*. Melbourne: Victorian Government Printer, June 1999, p 205.

"You will appreciate that count 1 is the hazard and count 2 is the risk, because the prosecution says it follows in a logical sequence. First of all, in a hazardous place like a gas plant you have to identify the hazards, that is the HAZOP, count 1; and then when you have identified the hazard, count 2, you have got to assess the amount of the risk, and it is a logical step. You first of all have to identify the hazard and then you have to assess the risk of it. That is why it is laid hazard first, risk second."

And a little later the Judge reiterated the point:

"There should have been an adequate periodic risk assessment done to assess the hazards which the HAZOP would have revealed."

These statements highlight the Court's confusion. It is true that, in the OHS formulation, risk assessment follows hazard identification. But Esso was not using the language of the OHS formulation. Its periodic risk assessments were *not* designed to assess the risks of hazards identified in HAZOPs. They were independent hazard identification procedures.

Risk management professionals appear able to live with this terminological confusion, but the Esso trial demonstrates that inconsistencies in the way "risk" is used can at times have serious consequences. The result, on this occasion, was a charge which simply did not make sense, a charge on which Esso was nevertheless convicted! It is clear that the language of risk management needs to be reassessed.

3. Failure to maintain plant in a safe condition

The third count was failure to maintain plant that was, so far as is practical, safe and without risk to health. This referred to the fact that, on the morning of 25 September, the metal heat exchanger became brittle with cold and therefore susceptible to cracking should it be subjected to a shock — either a physical shock such as being hit with a hammer, or a thermal shock such as the injection of warm oil. This was highly dangerous because the heat exchangers remained under pressure, and any fracturing of the vessel could be expected to result in the release of large quantities of highly flammable material. What Esso should have done as soon as the plant became cold was to isolate it and allow it to thaw out. Instead, operators tried to warm it up by slowly reintroducing warm oil, which shattered one of the heat exchangers.

This was the prosecution's case. Count 3 is hardly a controversial one — the obligation to maintain plant in a safe condition is a very clear legal requirement. However, Esso disputed the charge in a novel way. It admitted that the heat exchanger had become abnormally cold and perhaps even brittle, but it contended that this did not necessarily mean that the equipment was in a dangerous condition. It would only be dangerous if it was susceptible to failure and it would only be susceptible to failure if something else happened, for example, the reintroduction of warm oil. The argument is analogous to saying that an unguarded power saw is only dangerous if a worker puts his/her hand in it — not an argument which could be expected to carry much weight with a jury.

4. Inadequate procedures to deal with loss of warm oil circulation

The fourth charge was operating the gas plant without adequate procedures relating to the loss and restoration of warm oil circulation. As soon as the operators had ascertained that they were not going to be able to get the warm oil pumps restarted, there were certain steps they should have taken to ensure that plant did not become dangerously cold. They didn't take these steps because Esso had not specified procedures to be followed in these circumstances — in turn, because it had not identified this particular hazard.

Moreover, where procedures are critical, as they were in this case, it is important that they be documented. Thus, the essence of the charge was the absence of written procedures. The prosecution's position was that "you might not need written procedures in a milk bar if you spill some milk, but you do need them in a volatile and flammable gas plant with complex procedures".

The prosecution was not put strictly to the task of proving this charge because Esso made various admissions in the course of the trial. It did so because they were not matters which the company disputed and to have put the prosecution to the task of proving them would have wasted the Court's time. Thus, in relation to count 4 Esso admitted that:

> "There were no written procedures dealing with any ... hazards or dangers that may have existed associated with cold temperatures [in the area where the explosion occurred]."

5. Failure to adequately train employees to respond to a loss of warm oil

There was of course no training provided in what to do if the warm lean oil circulation failed because there were no procedures to be trained in. So it was that Esso was charged with failing to adequately train employees to respond to a loss of lean oil circulation. Again, the prosecution did not, in effect, have to prove this charge because of admissions which Esso made. Most relevantly here, Esso admitted that:

> "Operators and supervisors at the Longford plant were not trained with respect to and were not aware of whether or not cold temperature or hazards could arise in [the area of the plant where the explosion occurred] from the loss of lean oil circulation."

Esso had provided fairly generalised training for its operators.[12] But the training had not been of sufficient quality, nor had it been sufficiently focused on safety-critical procedures. The case thus provides useful guidance to companies on where to focus their training efforts. When hazards are identified, companies often specify that certain procedures be followed as part of their risk reduction strategy. Safety therefore depends on the adequacy of training which operators have received in these procedures. In short, it makes good sense to focus training efforts on procedures which are important for risk reduction purposes.

6. Failure to provide the means for ensuring that items of plant operated at a safe temperature

The essence of this charge was that Esso had not set critical operating parameters for the heat exchangers and related pieces of plant. Critical operating parameters are limits of temperature, pressure and so on, beyond which plant should not be operated. In this case there was no minimum temperature specified for the operation of the heat exchangers. Worse still, the temperature gauge on the unit only went down to zero and there was therefore no way of knowing how cold the unit was if it went below this point. It is only as a result of various calculations that investigators were able to estimate that the temperature dropped to -48°C.

7. Failure to adequately train employees regarding the risks associated with abnormally cold plant

This count sounds similar to count 5. However, the prosecution argued that they were different. Count 7 is the failure to train employees about the dangers of cold temperature, in particular the problem of embrittlement, and count 5 is the failure to train employees in how to respond to the problem.

Count 7 was laid as a separate charge because of its tragic consequences. Just minutes before the explosion, several employees were working on a leak which one of the heat exchangers had sprung as a result of becoming very cold. It was leaking at a rate of many litres per hour and the problem was seen by most of those present as a problem of environmental contamination, not as an indication of danger. All except one were oblivious to the danger they were in. That one person had previously been trained in the Navy about the dangers of embrittlement. According to the prosecution, this man was a living example of the importance of proper training. When he finished

12. Hopkins, A. *Lessons from Longford: the Esso gas plant explosion*. Sydney: CCH Australia Limited, 2000, p 18.

work on the leaking exchanger he declared, "Fuck this, I'm out of here", and left as quickly as possible. In so doing he saved his life. Two others died and several were injured because they remained at the scene. The general training which operators had received at Esso had touched on the issue of embrittlement and thermal shock, but it had not conveyed to them a clear idea of what thermal shock meant — nor how dangerous it was.[12]

8. Failure to adequately train supervisors regarding the risks associated with abnormally cold plant

Count 8 was the same as count 7, except that it concerns supervisors, not rank and file employees. Supervisors were equally untrained in the dangers of cold plant and hence equally ignorant. They were therefore not in a position to provide appropriate guidance to operators and maintenance workers in the vicinity. Both of the men who died at the scene were in supervisory positions.[13]

The real significance of this charge is that it counters some of the implicit criticism of plant operators which runs through Esso's account of the incident. A letter from Esso's solicitors noted that the workers had failed to get help in dealing with the situation, implying that they should have. However, had they sought help, it would have been from supervisors who were equally unaware of the danger.

9. Failure to monitor conditions

The essence of this charge is that Esso failed to monitor conditions at the plant by failing to provide properly functioning instruments and on-site engineers. A variety of temperature, pressure and flow recorders at the plant were supposed to provide an indication of how these measures varied over time. In theory, this enabled operators to monitor trends, in the way that instruments which provide only the current readings do not. As the prosecution put it: you need a continuous record so that you can see where you have come from in order to know where you are going. Most of these recorders used paper and ink to provide a continuous readout, and many of them were not working, either because they had no paper in them or because the pens were not working.

Esso attempted to shift the blame for this to an operator whose job it was to keep the paper and ink flowing:

> "Esso did provide a system for the charts to run, a very simple one, and one that could be operated by anyone and surely, we say to you, that Esso can't be guilty of a crime in that respect, because it didn't have a supervisor or valet to make sure that Mr ... did these simple tasks."

The operator explained, however, that while maintaining the recording charts had been a company priority in the 1980s, by 1998 it was not. Perhaps the most telling point in his favour is that Esso's engineers were clearly not using these charts; had they been doing so they would have realised that very little recording was taking place.

The second part of the charge concerned the absence of on-site engineers. Esso had had engineers on site at Longford to monitor conditions until the early 1990s. At that time, in order to cut costs, it withdrew its engineers to its Melbourne headquarters and expected them to monitor the plant from afar. However, much of the instrumentation, in particular the fact that trend data were only available in hard copy at Longford, did not facilitate monitoring from afar.

The prosecution conceded that a fully computerised system which was capable of displaying trend data on computers at head office might have made it feasible to monitor the plant from Melbourne but, given the nature of the instrumentation in use at the time, it was essential that engineers be on site.

It should be noted that the Royal Commission did not concede the point conceded by the prosecution:

> "The physical isolation of engineers from the plant deprived operations personnel of engineering expertise and knowledge which

13. Longford Royal Commission Report, op cit, p 36.

previously they had gained through interaction and involvement with engineers on site. Moreover, the engineers themselves no longer gained an intimate knowledge of plant activities. The ability to telephone engineers if necessary, or to speak with them during site visits, did not provide the same opportunities for information exchange between the two groups, which are often the means of transfer of vital information."[14]

An expert witness who gave evidence at the trial expressed a similar view. Had engineers been on site on that Friday morning when things had begun to go seriously wrong, and had they seen the ice on the vessels which are normally too hot to touch, they would have recognised the danger and cautioned against reintroducing hot oil.

Esso has since admitted that it did not have plant operations effectively under surveillance by its engineers and has installed systems which enable effective monitoring of safety-critical indicators from head office. As a result, process upsets are not now left entirely to operating staff to deal with, as they previously were.

10. Failure to ensure that the heat exchanger was protected from thermal shock

Some time after the heat exchangers had become brittle with cold, operators succeeded in getting the warm oil pumps going, thereby reintroducing hot oil into the dangerously cold vessel. This critical event was referred to at the trial as "hot hit cold" or "hot on cold". The result was a thermal shock which ruptured the vessel. It would have been quite practicable for Esso to have installed a temperature sensing mechanism and an automatic trip to stop hot oil from entering dangerously cold vessels. This would have prevented the operators from making the mistake they did. Such a trip mechanism was the last in the chain of protective mechanisms and procedures which should have been in place, that is, proper instrumentation, surveillance, training, and so on. It represented the last line of defence which would have protected the plant against the incident, should all other defences fail. This is the philosophy of defence in depth, which is widely accepted in hazardous industries.[15]

The logic of this charge would appear to be the same as all of the others in the sense that they all refer to the absence of preventive action which Esso should have taken. It was the failure to take this preventive action which was the offence, not the fact that an incident actually occurred or that people were killed or injured. Theoretically, even if no incident had actually occurred, the company could have been charged and convicted of all of these offences.

However, the prosecution treated this charge as very different to the others. It declared that the failure to have such a trip mechanism in place was only an offence because the incident had actually occurred, that is, because hot oil entered the cold vessel and ruptured it.[16] The prosecution's reasoning was never satisfactorily explained, leading the Judge at one point to make the following comment to Esso's counsel:

> "I must say, no doubt through my own deficiencies, I don't even know why it needs to be proved for count 10 that hot lean oil entered, but the prosecution says it does so I thought I'd give you that benefit as well."

This is an intriguing comment. The Judge clearly thought that all the prosecution needed to prove was that a trip mechanism would have been a practicable preventive measure. However, the prosecution was saying that it also needed to prove that hot oil had actually entered the cold vessel. The result was that the prosecution was making things harder for itself than the Judge thought necessary. But, by making things harder for itself, the prosecution was making things easier for the defence, and since in a criminal

14. Longford Royal Commission Report, op cit, p 209.

15. Reason, J. *Managing the risks of organisational accidents*. Aldershot: Ashgate, 1997.

16. Counsel told the jury: "We actually say [to Esso] in count 10, 'Look, you blew it up by putting hot lean oil into it and you should have had a fail-safe mechanism to stop the hot lean oil getting into a thermally cold vessel that became shocked'."

trial the defendant must be given the benefit of every doubt, the Judge chose not to argue the matter with the prosecution.

Esso argued that most of the other counts similarly required "hot on cold", that is, they were only offences if it could be shown that hot oil actually entered the cold heat exchanger causing it to rupture, but the Judge rejected these arguments on the ground that the counts were about the *potential* for disaster not its occurrence.

Esso never conceded that hot oil ever entered the heat exchanger and the prosecution was therefore forced to argue at great length that this is what in fact happened. Indeed, counsel for the prosecution complained at one stage that the trial had been "highjacked" by the question of whether hot hit cold. "It is important not to be side-tracked into the notion that causation is the heart of the case", he said. Counts 1 to 9 did not allege that the explosion had been caused by hot hitting cold. They simply alleged that Esso had not identified and controlled the risk of hot hitting cold. Esso would be guilty as charged even if it could be shown that hot had not hit cold on this occasion, and that the rupture had been caused by a meteorite hit. On the other hand, he said, count 10 did depend on establishing that hot hit cold. Had the prosecution not insisted on this point, it could have saved itself considerable effort.

There is one significant consequence of this prosecution strategy. It was noted earlier that when juries find a defendant guilty, they do not provide their reasons and there is no way of knowing which of the peripheral claims the prosecution may have made along the way were accepted by the jury. If "hot on cold" had not been specified as part of count 10, there would have been no way of knowing at the conclusion of the trial whether the jury accepted that hot oil had actually entered the heat exchanger, causing it to rupture, or whether the rupture had occurred in some other way, as suggested by the defence. But because "hot on cold" was specified as part of count 10 and because the jury found Esso guilty on this count, it can be concluded that the jury indeed found that "hot on cold" was the immediate cause of the rupture. In short, although the trial was about Esso's failure to take adequate precautions and not about whether this failure had caused the explosion, the jury ended up returning a verdict on both of these matters. It is worth noting that these findings were in agreement with those of the Royal Commission. The full significance of the jury's finding about causation will be dealt with in the next chapter.

11. Failure to ensure the safety of persons other than employees

This count referred to the fact that police, ambulance workers and Country Fire Authority (CFA) personnel were placed at risk by having to attend the disaster scene. The count was similar to some of the earlier counts but it differed in that those counts concerned placing employees at risk; this count concerned placing outsiders at risk. It was based on section 22 of the OHS Act which places an obligation on employers to ensure, so far as is practicable, that their operations do not place at risk persons who are not employees.

Prior to the trial, Esso sought to have this charge dismissed on the grounds that it had no control over emergency workers once they had decided to come onto the premises, and it was not practicable for Esso to have prevented these people from being exposed to danger. It submitted that perhaps the CFA was liable under the OHS Act for exposing its workers to risk.

The Judge was scathing about Esso's attempt to have count 11 dismissed: "The submission I have just had the misfortune to hear is both wrong in law and callous in character."[17]

His reasoning as to why it was wrong in law was as follows. If the fire had not occurred, emergency workers would not have been exposed to risk. If the evidence showed that it was practicable for Esso to have prevented the risk to its own workers, then it would follow automatically that it would have been practicable to prevent the risk to emergency workers. On the other hand, it was not practicable for the CFA

17. Ruling no 14, 30 May 2001.

to avoid exposing its workers to risk when they attended a fire — the risk was inherent in the job.

As to why the submission was callous, the Judge noted that:

> "In order to tender care and succour and protection to Esso's personnel and property, decent Australians — police, ambulance and fire personnel — attended on the afternoon of 25 September 1998 and put their lives and safety at risk."

He went on to describe in graphic terms the dangers which these workers had faced. And yet Esso claimed that it had not exposed emergency personnel to risk. It was this claim which incensed the Judge. Needless to say, having convicted Esso on all other counts, the jury convicted Esso on count 11.

Practicability and foreseeability

The OHS Act imposes a duty on employers to provide a safe workplace "so far as is practicable". The mere fact that the workplace is dangerous does not mean that an employer has committed an offence; it is only if the employer has not done what is practicable to reduce or eliminate the risk that it would be liable. In particular, the fact that someone may have been killed does not in itself mean that an employer has failed in its duty; it is only if the death was practicably preventable that an employer may be found guilty.

Moreover, in Victorian law it is up to the prosecution to prove beyond reasonable doubt that it was practicable for the employer to have controlled the risk in question.[18] Thus, in all of the preceding counts the prosecution had to prove not only that Esso failed to take certain action, but that it was practicable for Esso to have taken that action — identify the hazard, control the risk, train its personnel, and so on. In addition, the jury had to be convinced of this. Accordingly, the Judge gave instructions to the jury about the meaning of practicability at law:

> "In deciding what was practicable, you should have regard to:
>
> (a) the severity of the hazard or risk in question;
>
> (b) the state of knowledge which the employer had or ought to have had about that hazard or risk and about any ways of removing or mitigating that hazard or risk;
>
> (c) the availability and suitability of ways to remove or mitigate that hazard or risk; and
>
> (d) the cost of removing or mitigating that hazard or risk."[19]

This was not only the legal meaning of practicability, he said; it was "really just commonsense".

In relation to (a), the Judge explained that the risks in a sawmill were far more severe than those in a library, and that employers could be expected to go to greater efforts to control risks in the former than the latter. There was no dispute that a gas processing plant was a high-risk environment.

Further, he told the jury that (c) and (d) were not at issue, in that there were obviously ways in which the risks could have been mitigated, ways which were well within the financial capabilities of a company like Esso.

Point (b), which concerned the state of knowledge of the employer, was the one about which the Judge had most to say. It was not the knowledge which the employer actually had about the risks and ways of mitigating them which was most relevant, it was the knowledge the employer *ought* to have had:

> "Some companies may be appallingly slack and it's no use for the company to say, 'We didn't know about it, sorry, because we had our eyes shut and were looking the other way.' It's not a subjective standard; it's an objective standard."

18. The situation is similar in most other States and Territories. In NSW, it is for the defendant to prove on the balance of probabilities that it was not reasonably practicable to avoid the risk in order to escape conviction. See Johnstone, R. *Occupational health and safety law and policy*. Sydney: Lawbook Company, 1997, ch 5.

19. This is in fact a restatement of section 4 of the OHS Act.

He explained further that in making judgments about what the company ought to have known, the jury should not be relying on hindsight — "we are all wise after the event". The question was one of foresight — should the company have foreseen the possibility of embrittlement and rupture? The Judge noted that the prosecution argued that it "was all plainly and obviously foreseeable" and he quoted some of the evidence on which the prosecution relied. Counsel for the prosecution had had the following interchange with an expert witness:

> "Q: In your opinion, how foreseeable was the possibility of cold developing down at the hot end of the plant; was it foreseeable; is it something that is foreseeable to a designer, to a company or entity operating a plant like this?
>
> A: Yes, it is foreseeable.
>
> Q: What makes it foreseeable?
>
> A: You are dealing with high pressure fluids, liquids and gases and if you have a pressure reduction they will get cold. They are already cold to start with and they will get colder. If they are allowed to flow into an area of plant which is not designed for low temperatures then the potential for brittle failure ... is foreseeable."

The main question for the jury, then, in judging whether it was practicable for the company to have taken the various preventive actions specified in the 11 counts was whether the possibility of hot on cold was foreseeable. In returning guilty verdicts on all 11 counts the jury clearly decided that it was.

The issue of multiple counts

In general, the criminal law requires that, if the prosecution is alleging multiple offences, each must be the subject of a separate count. In this way the accused knows exactly what is being alleged and can organise an appropriate defence. Where a single count covers more than one allegation, it is said to be "duplicitous" and therefore unacceptable. A Victorian Supreme Court ruling has emphasised the importance of this principle in the case of prosecutions brought under the OHS Act and so in that State companies are often charged with multiple counts.[20,21]

While multiple counts may avoid the problem of duplicity, their use raises another problem, namely, the possibility that some of the counts may be so similar that the defendant is in effect being charged more than once for the same offence. This is the problem of repetition.

When Esso first appeared in the Magistrates' Court, it was charged with a total of 45 offences — presumably to avoid duplicity.[22] By the time the case first appeared in the Supreme Court, these had been reduced to 21 — presumably to avoid repetition. The Judge considered these counts and decided that "there was still a significant degree of unnecessary charging".[23] For example, the OHS Act at one point uses the phrase "information, instruction [and] training" and the prosecution had made each of these the basis of a separate count. The Judge decided that information, instruction and training were effectively the same thing, and the prosecution thereupon chose to focus on training and to drop counts relating to information and instruction. In the end the Judge approved 11 counts and ruled that there was no repetition, duplication or unfairness involved.[24]

There is clearly scope for disagreement with this ruling and Esso's counsel disagreed vehemently. "The inequity and repetitive nature of these charges is nowhere better demonstrated than with these training counts", he said. He pointed out that supervisors are

20. Reasons for Ruling no 5, 13 February 2001, para 21; Johnstone, op cit, pp 191ff.
21. The situation in NSW is not so clear. See Thompson, W. *Understanding NSW occupational health and safety legislation*. Sydney: CCH Australia Limited, 2001, pp 29-30.
22. Reasons for Ruling no 5, 13 February 2001.
23. Ruling no 5, 13 February 2001, para 4.
24. Ibid, para 8.

just as much employees as are operators and that it is repetitive to charge Esso with failing to train both employees and supervisors. Likewise, the failure to ensure the safety of employees seems similar to failure to ensure the safety of non-employees. There is obviously a considerable degree of repetition involved in these counts.

But there is another more interesting way in which the charges overlap. There is an important sense in which many of these failures follow one from the other. The initial count was failure to carry out an appropriate hazard identification process. Had a HAZOP been done, it would almost certainly have identified the problem of embrittlement. A risk assessment (in the OHS sense[25]) would have been carried out automatically (count 2). Procedures would have been developed to deal with the cessation of warm oil flow (count 4) and operators would very likely have had training in those procedures (count 5). Esso would probably have specified critical parameters for the operation of the heat exchangers (count 6). The plant would therefore not have ended up in a dangerous condition on the morning of the incident (count 3). Automatic trips to protect super cold vessels against warm oil would very likely have been installed (count 10), and emergency workers would not have been exposed to danger (count 11). In short, had the first offence not occurred few, if any, of the others would have. Conversely, most, if not all, of the offences followed almost inevitably from the first.

The Judge was well aware of the sequential nature of the counts. He described them at one point as steps along the same path. The prosecution used an even more dramatic metaphor:

> "Once you get over the first charge, the rest of it, leaving aside for the moment counts 10 and 11, the rest of it goes like a pack of cards, almost ... The logic of the situation is pretty remorseless."

It might appear from this analysis that it was unfair to treat each of these matters as a separate offence. But despite the remorseless logic, the prosecution was insistent, and the Judge agreed, that the charges were independent and each needed to be considered on its merits. The impression remains, however, that in bending over backwards to avoid duplicity, the Court was substituting one kind of unfairness for another.[26]

One important consequence of the multiple count strategy deserves noting. The maximum possible fine under the OHS Act is $250,000, plus an additional $250,000 if the offence is not the first, making a total of $500,000. Had Esso been charged with a single count, this is the maximum it could have been fined. But because it was convicted on 11 counts, the combined penalty was far higher — $2m — a mega-fine in comparison with other fines which have been imposed in Australia for OHS offences.

Although the multi-count strategy resulted in the imposition of a mega-fine, there has been no suggestion that that was its purpose. In contrast, in the US the Occupational Safety and Health Administration has at times deliberately resorted to a multi-count strategy in order to obtain a fine which was commensurate with the severity of the offence. A building collapse in the US in 1987 resulted in the deaths of 28 construction workers. Inspectors found that 238 metal brackets used to support a concrete floor were below required strength and the Agency imposed the maximum penalty of $10,000 in each instance, totalling $2.38m.[27]

One wonders whether the Esso case will encourage Australian OHS agencies in the future to make use of multiple counts in order to obtain mega-fines in circumstances where they are thought to be appropriate.

25. See the discussion under count 2 above.

26. This possible unfairness was taken into account at the time of sentencing. See ch 4.

27. *Corporate Crime Reporter*, 26 October 1987, p 2. Occupational Safety and Health Administration fines are "proposed" and then negotiated downwards.

Conclusion

The number of charges brought against Esso suggests that the prosecution "threw the book" at the company. But this does not seem to have been a deliberate strategy aimed at maximising the fine. Instead, the prosecution's aim was to dissect the incident and identify the many distinct ways in which Esso had failed to provide a safe workplace. One consequence of this was to maximise the learning opportunities — for other companies as well as for Esso. There was undoubtedly some duplication, and one charge was rendered meaningless by the Court's confusion between risk and hazard. But the prosecution's strategy did provide a systematic approach to the multiple failures which contributed to the explosion. In particular, it demonstrates that courts are willing to take the sequence of hazard identification, risk assessment and risk control as fundamental and to identify and prosecute failures along this chain. The prosecution did not rely on any Regulations which may have required this approach. It argued, and the jury agreed, that this was implied in the fundamental duty of care. Moreover, the Court reinforced the importance of training, not just any training, but training identified as necessary for the control of specific hazards. Given that the Court has taken the risk management approach so seriously, safety professionals should as a matter of urgency clarify the terminological ambiguities.

CHAPTER 3

Esso's defence

The preceding chapter dealt largely with the prosecution's case. Not much was said about the way that Esso defended itself. In this chapter we look in more detail at Esso's defence. This is a matter of interest in its own right — but that is not the main reason for discussing Esso's defence strategy. It turns out that the way Esso conducted its defence had significant consequences for the sentence which it received. We therefore need to examine Esso's defence strategy in order to understand how the Court assessed Esso's culpability.

In the traditional criminal law, two things must be established in order to establish the guilt of a defendant: (1) that the person (or company) did the prohibited deed, or failed to do what ought to have been done; and (2) that the person had a criminal state of mind, normally, "an evil intention, or a knowledge of the wrongfulness of the act".[1] Corporate failure to ensure the health and safety of workers seldom involves an evil intention or knowledge of the wrongfulness of the act. To require the prosecution to prove a criminal state of mind in this sense would make employers virtually immune from prosecution. Accordingly, OHS statutes do not require the proof of *mens rea*, to use the legal term.[2] Occupational health and safety offences are strict or absolute liability, in the sense that the employer's state of mind is irrelevant in deciding whether or not an offence has occurred.[3]

It is sometimes argued that because strict or absolute liability offences do not require the proof of a criminal state of mind, they "are not criminal in any real sense, but are acts which in the public interest are prohibited under penalty".[4]

The Esso trial, however, was explicitly a criminal trial. It took place in the criminal division of the Supreme Court and the Judge made it clear that Esso was charged with criminal offences.[5] Esso would be given all the protections available to criminal defendants, for instance, the right to know clearly just what is being alleged and the requirement that the prosecution prove its case beyond reasonable doubt.[6]

1. Johnstone, R. *Occupational health and safety law and policy*. Sydney: Lawbook Company, 1997, p 200.
2. Ibid, pp 199-202.
3. The terms "absolute liability" and "strict liability" are often used interchangeably in OHS matters. If a distinction is made, as in *He Kaw Teh* ((1984) 157 CLR 523), then the liability of employers under OHS statutes is absolute, rather than strict. See Thompson, W. *Understanding NSW occupational health and safety legislation*. Sydney: CCH Australia Limited, 2001, pp 25-27.
4. Johnstone, op cit, p 200.
5. Ruling no 2, 16 August 2000, para 2, 8.
6. Ruling no 2, 16 August 2000.

This had important consequences for the conduct of Esso's defence. It meant that Esso did not have to prove that it was innocent of the charges and it did not have to offer any explanation for how the explosion took place. It was entitled simply to raise doubts about the prosecution's case and then to argue to the jury that the prosecution had not proved its case beyond reasonable doubt.

Questioning the causal connection

The principal strategy which Esso used was to question the prosecution's claims about what caused the heat exchanger to rupture. In so doing it was also questioning the conclusions of the Royal Commission as to the causes of the explosion. The prosecution argued, it will be recalled, that the loss of warm oil caused the heat exchanger to become brittle with cold and the reintroduction of warm oil caused it to rupture. The defence sought to question the extent and relevance of embrittlement and even whether the operators actually succeeded in reintroducing lean oil into the heat exchanger.

In relation to embrittlement, the defence argued that metalurgical evidence showed that the rupture had started at a defect in one of the welds in the vessel and that, had it not been for this defect, the admittedly brittle vessel might not have shattered:

> "This fracture occurred only because of the defect in the weld, a defect which, on the evidence, was unknown to Esso, one that had existed despite construction by an apparently reputable manufacturer ... [and] inspection during the time of fabrication in accordance with what was required in those days and inspection since that time."

Embrittlement, in other words, was not the critical issue which the prosecution had made it out to be. The weld defect was. And this was a matter about which Esso had no way of knowing and for which Esso could not be held responsible.

This argument was easily countered by the prosecution. If there is to be a rupture, it will naturally start at a weak point in a vessel. Moreover, there are always tiny defects in welds which can provide a starting point for a rupture. This is quite foreseeable and construction codes generally assume a certain level of defects in a vessel when calculating safe minimum temperatures. Accordingly, the prosecution argued that "to point to the weld is not a defence, it is not an excuse; it is no more than an historical explanation of how the rupture propagated itself".

Probably more vital from Esso's point of view was its claim that the prosecution had not established that hot oil actually entered the embrittled vessel. It suggested that the lack of any visible melting of ice on the inlet pipes after warm oil had allegedly begun flowing suggested that the warm oil had not reached the heat exchanger by the time of the rupture. And it argued that when one of the operators opened another valve just 16 seconds before the rupture, this might have provided a pressure shock — nothing to do with hot on cold — which caused the embrittled vessel to rupture.

The Royal Commission had explicitly rejected the idea that the valve which was opened 16 seconds prior to the rupture had anything to do with it, and counsel for the prosecution suggested that this theory was about as fanciful as suggesting that the rupture might have been caused by a meteorite hit.[7]

Of course, it was not up to Esso to demonstrate the truth of its alternative hypotheses about what might have caused the rupture. The onus was on the prosecution to rule out any plausible alternatives. If it could not, then it could not claim to have established the causal sequence beyond reasonable doubt. Esso's strategy therefore was, first, to try to raise doubt about the prosecution version, and second, to suggest alternative hypotheses.

Indeed, at one point Esso wrote a letter to various experts around the world asking them if they could come up with alternative hypotheses about how the

7. Dawson, D and Brooks, B. *The Esso Longford gas plant accident: report of the Longford Royal Commission*. Melbourne: Victorian Government Printer, June 1999, p 95.

rupture might have occurred. The purpose of this exercise was presumably to cast doubt on the prosecution's case.

Esso's position was that if a cause could not be conclusively established, then the Court would have no alternative but to regard the rupture and explosion as an accident. As Esso's counsel rather plaintively put it:

> "Why can't it be that you can have an accident in this world of ours? Why can't it be that you do not know how something happened?"

As we shall see, these words came back to haunt Esso at the time of sentencing.

All of this raises an important question. Why was Esso so interested in casting doubt on the prosecution's theory of what caused the incident? After all, causation was not the heart of the prosecution's case. The counts were mostly about the failure to take practicable precautions. Whether or not these failures were responsible for the rupture which occurred on this occasion was not the point; they might have been. It was the failure to take precautions which was the offence, not the outcome of the failure to take precautions.

One can only speculate about Esso's sustained attack on the theory of hot on cold. One possible explanation is as follows. It was fairly obvious that Esso had failed to take practicable precautions. However, if Esso could convince the jury that it would only be guilty on these counts if the failures had actually caused the explosion, then it could attempt to defend itself by questioning the causal connections. If this was indeed Esso's game plan, it explains two puzzles: (1) Esso's otherwise inexplicable insistence that the prosecution needed to show causation in order to gain convictions; and (2) Esso's dogged attempt to undermine the prosecution's causal explanation.

There was another motive for undermining the theory of hot on cold, a motive which went way beyond the circumstances of the criminal trial. What was at stake in the criminal trial was insignificant in comparison with what was at stake in the looming civil case. The Longford disaster interrupted the supply of gas to Victorian industry for two weeks and these industries (or their insurers) were suing Esso for $1.4b. This was many orders of magnitude more than any fine which might result from a criminal conviction. It seems likely that Esso would have defended itself at the criminal trial with an eye to this main game.

One possible basis of the civil action against Esso is negligence. In order for a negligence action to succeed, the person harmed must establish both that the person sued was negligent and that the negligence caused the harm. It would be hard to avoid a finding of negligence, since negligence means the failure to take practicable precautions, and it would be difficult to persuade a civil jury that Esso had taken all practicable precautions. But if Esso could undermine the theory of hot on cold, it might persuade a jury that whatever caused the explosion it was not Esso's negligence. In this way it might escape liability. If this was indeed Esso's reasoning, it is very clear why the counsel for Esso in the criminal trial devoted so much effort to testing the prosecution's theory of what caused the explosion.

In the end, as we saw in the previous chapter, the conviction on count 10 was in effect a finding by the jury that hot had hit cold and that this was the cause of the explosion. This was a significant setback for Esso in its fight to avoid civil liability.[8]

Esso's position on cause

What, then, was Esso's position at the criminal trial on the cause of the rupture? It was not saying that the company itself was in the dark about the causes. As counsel put it, "I didn't say that Esso didn't know".

8. Shortly after the sentence was imposed, the issue was raised in preliminary court proceedings in the civil action. The Judge in this matter noted the findings in the Royal Commission and at the criminal trial and suggested that they supported an inference of negligence. He asked Esso's lawyers whether they would be willing to admit negligence. If Esso made such an admission, he said, the civil trial would not need to explore factual issues about the cause of the explosion, which would result in a dramatic saving of time. *The Age*, 17 August 2001, p 9.

Its position was simply that the Court and the jury couldn't know because the evidence presented by the prosecution was insufficient. And Esso was under no obligation to provide any evidence to the Court — that was for the prosecution.

Despite Esso's position that it was not saying what it knew or didn't know, the Judge noted that several of counsel's statements appeared "subliminally" to go beyond this and suggest that Esso really *didn't* know what caused the rupture: for example, "Esso hadn't made any conclusions whatsoever", and "why can't it be that you do not know how something has happened?"

One can imagine various reasons why Esso may have wished to convey the impression that it did not know. First, if Esso, with all its resources, really didn't know what caused the rupture, even after the event, then perhaps it was after all an accident for which Esso could not be held responsible. Second, if the jury formed the impression that Esso had come to its own conclusions about the causes of the rupture but was not letting on, this might influence the jury against Esso. From this point of view it would be best for Esso if the jury had the impression that Esso really did not know.

But what concerned the Judge about this subliminal suggestion that Esso did not know was that the company had produced no evidence that it had failed to identify the causes of the rupture. For counsel even to suggest that the cause was a mystery to Esso was therefore going beyond the evidence.

There was in fact powerful evidence produced late in the trial that Esso had done its own investigation which, to a large extent, supported the prosecution's position. The earlier-mentioned letter which Esso sent to various experts asking for alternative hypotheses was discovered by the prosecution late in the trial. In setting the scene for the request, the letter acknowledged that the loss of warm oil circulation had caused the heat exchanger to become brittle with cold. It bluntly stated that "the failure was due to embrittlement following the vessel being subjected to temperatures below those for which its metal was designed". This was something which Esso did not concede at the trial and which it forced the prosecution to go to great lengths to prove. The letter did not admit the last element in the causal chain — that the stressor which triggered the rupture was the introduction of hot oil — but the rest of the explanation was there, causing the prosecution to call the letter a "smoking gun", that is, a conclusive piece of evidence that Esso knew and conceded almost all of what the prosecution had alleged about the causes of the rupture. In passing sentence later, the Judge commented on this situation as follows:

> "Esso sought to make it appear that the identification of hazard, risk and cause was impossibly difficult. To that end, prosecution experts were cross-examined in technical detail to undermine proof which, with its other hand, Esso was promoting to its own experts."[9]

Would the HAZOP have revealed the hazard of embrittlement?

Another major aspect of Esso's defence was to question whether a HAZOP of Gas Plant 1 would have identified the problem of embrittlement. If Esso could successfully argue that a HAZOP would have missed this problem, it could then argue that the rupture was not foreseeable and therefore not preventable. Esso tried to make this argument in two ways.

First, Esso produced an expert witness and put to him a question about whether the HAZOP might have been expected to identify the sequence of events which caused the rupture. It will be remembered that the sequence of events leading to the loss of lean oil began the night before and was a long and intricate chain of events. Counsel for Esso put to the witness the whole incident sequence — the events preceding the shutdown of the warm oil pumps as well as those following — and asked him whether a HAZOP would

9. The sentence, para 45 (see Appendix at p 65).

have been likely to identify this sequence. So complex was the entire chain of events leading to the embrittlement of the heat exchanger that the question took up 38 lines of transcript. Predictably, the expert's view was that a HAZOP would not have identified this entire sequence of events.

But that is not the function of a HAZOP. A HAZOP focuses on one point in the system at a time and asks "what if?" questions: what if the temperature here was too high, too low, etc? The question which counsel for Esso asked its own expert was really a red herring.

Later, at the time of sentencing, the Judge was extremely critical of this question:

> "The convoluted and obscure question asked by the defence of the defence witness... — all 38 lines of it — and the convoluted and obscure scenario posited by [the defence witness] ...— all 58 lines of it — are testimony to the defence obfuscation; and the words were hollow when [counsel for the defence said of the question] ... 'That's about as simply as I can put it', ... [and the defence witness said] 'I'll try and keep it as simple as I can'."[10]

However, the fact that Esso had placed its own expert in the witness box to give evidence meant that this witness could now be cross-examined by the prosecution. The prosecution jumped at this opportunity and, as was noted in the previous chapter, asked the witness to carry out an imaginery HAZOP of the relevant part of the plant. He did so and was forced to agree with the prosecution that the HAZOP team would see the problem of embrittlement "coming at them like a steam train". Here, then, was Esso's own witness admitting that a HAZOP would certainly have identified the problem of embrittlement. Esso's strategy had entirely backfired.

How had this happened? Esso was well aware that it was dangerous to put its own witnesses in the witness box precisely because this would expose them to cross-examination. They might then be forced to make admissions against Esso and to reveal what Esso knew about the causes of the explosion. For this reason its game plan was to produce no evidence of its own but simply to contest the prosecution's evidence. But on the question of the HAZOP, Esso apparently decided that it was worth taking the risk of putting a witness in the box to make the point that no HAZOP could have identified the entire incident sequence. Unfortunately for Esso, its gamble did not pay off and counsel for the prosecution succeeded in turning Esso's witness into a star witness for the prosecution.

It is ironic that the two most dramatic pieces of evidence in the whole trial turned up by chance. These were: the letter in which Esso admitted most of the prosecution's case as to cause ("the smoking gun") and the Esso witness who became a witness for the prosecution by admitting that a HAZOP would have identified the problem of embrittlement. Neither of these was available to the prosecution as the trial began. This is not to say that the prosecution would have failed without them, but in the minds of the jury these two pieces of evidence must have put the case beyond all doubt. As far as the Judge was concerned, they demonstrated that much of Esso's defence was reprehensible. Referring to Esso's strategy in the trial he observed that "the defence advanced was one of obfuscation — designed not to clarify, but to obscure".

The second way in which Esso tried to undermine the prosecution's case in relation to HAZOPs can be dealt with more briefly. If the company could show that similar plants around the world which had done HAZOPs had not identified embrittlement as a problem, this would cast doubt on whether a HAZOP at Gas Plant 1 at Longford would have identified the problem. To this end, Esso selected 23 plants in the US which had done HAZOPs and not come up with the problem of embrittlement. It hired a consultant, escorted him around these plants and then called him as a witness. Questioned by the defence counsel the witness expressed the view that the plants were essentially similar to Gas Plant 1 at Longford.

10. The sentence, para 45 (see Appendix at p 65).

However, under questioning from the prosecution he acknowledged a variety of differences between Longford and the plants he had seen. The prosecution's view was that the expert had been kept in the dark about how the 23 plants were selected and that "what he was being shown was not comparing like with like; not comparing apples with apples", as the Judge put it in his summing up. If so, the fact that their HAZOPs had not come up with embrittlement as an issue proved nothing.

There is an independent reason to doubt that these plants were equivalent to the Longford plant. Five were owned by Exxon and had presumably received the same instructions as Esso Australia about the need to be especially on the lookout for the very thing which happened at Longford. If so, their failure to pinpoint the problem of embrittlement suggests a failure to attend to Exxon's instructions, rather than a limitation of the HAZOP method. Either they are all guilty of the same offence as Esso, which seems improbable, or they are not plants where embrittlement could occur in the way it occurred at Longford.

Blaming the operators?

A final interesting feature of Esso's defence was its treatment of the principal operator involved in the attempt to re-establish the flow of warm oil. Esso had blamed this man at the Royal Commission: he "was in possession of the necessary information" and had failed to take appropriate action "due to reasons peculiar to himself".[11] Public opinion following this submission had been overwhelmingly against Esso. No less a person than the State Premier expressed the view that Esso's submission was stupid and that it had assassinated the character of one of the workers.

The same man appeared at the trial as a prosecution witness. If Esso's counsel cross-examined him there was a high risk that the jury would see this as an attempt to blame him again. This was a risk which Esso could not afford to take. It therefore decided not to cross-examine this witness at all.

The Judge was critical of every aspect of Esso's behaviour with respect to this man. Had Esso blamed this operator at the Royal Commission, he asked, and if so, was "the conduct of Esso cynical"? And again, by not cross-examining the operator at the trial, had Esso done an about-face, and if so, was this in itself "cynical conduct"? Esso answered no to all of these questions. But even though Esso had made a tactical decision not to blame this operator, counsel for the defence apparently could not help himself and blamed him on at least one occasion during the trial — for failing to keep the recording charts in working order.

The Judge was clearly not satisfied with Esso's denials and in his sentencing remarks he went to great lengths to say that this operator and the other employees who had appeared in the witness box were "brave, decent and impressive men".[12] He went on:

> "The truth is there was only one entity responsible for lack of knowledge on that day: Esso. It, and it alone, should have properly trained the operators and supervisors not only in production, which it did, but also in safety. It, and it alone, failed to do so. [The operator] and the employees did not fail. Esso failed."

Conclusion

It is clear that the tactics which Esso employed for its defence were a failure. This conclusion is not based on the fact that Esso was found guilty. The preceding analysis shows in some detail how the decision to plead not guilty, and to force the prosecution to prove details of the case which in the end were indisputable, backfired in a number of ways. The prosecution simply outmanoeuvred Esso. But, more importantly, the defence strategy backfired by irritating the Judge. This had profound consequences when it came to sentencing.

11. Hopkins, A. *Lessons from Longford: the Esso gas plant explosion*. Sydney: CCH Australia Limited, 2000, p 12.

12. The sentence, para 44 (see Appendix at p 65).

CHAPTER 4

Sentencing

Following Esso's conviction by the jury, it was now for the Judge to pass sentence. Not only did he determine a sentence for each count (see Table 1 on the following page), but he also provided a detailed statement of his reasons (see Appendix at p 59).[1] Given that decisions in the Supreme Court are authoritative, lower courts can be expected to follow the Judge's lead. The sentencing principles enunciated in the Esso trial therefore have wide ramifications.

The structure of the Judge's statement is significant. After some preliminaries about the provisions of the *Occupational Health and Safety Act 1985*, the statement begins with the following pronouncement:

> "What occurred at Gas Plant One at Longford on 25 September 1998 was no mere accident. To use the term 'accident' denotes a lack of understanding of responsibility and a lack of understanding of cause."

The Judge then quoted with approval the words of a Victorian Government Minister (who was at one stage responsible for OHS):

> "Too often one hears the response, 'But that was an industrial accident'. This carries a connotation of inevitability, which denies the possibility of prevention. Even worse, it implies that an offence that results in a work-related fatality is not as serious as other criminal offences involving fatalities."

The Judge went on:

> "The events of 25 September 1998 were the responsibility of Esso; no one else. Their cause was grievous, foreseeable and avoidable. Their consequence was grievous, tragic and avoidable."

Here at the outset was the Judge's own repudiation of the defence which Esso had advanced at the trial. The company had said it was an accident; the Judge said it was not. Esso had said the incident was unpredictable; the Judge said it was quite foreseeable. Esso had said what happened was unavoidable; the Judge disagreed. Esso at times attributed some responsibility for the incident to its operators; the Judge said responsibility was Esso's and no one else's. Finally, in the Judge's view, Esso's offence was "grievous", that is, culpable or blameworthy — so serious in fact that on two counts he imposed the maximum possible sentence of $250,000. He hinted that he would have imposed higher penalties had he been able to.[2] It is rare for courts to impose the maximum possible penalty allowed under legislation. The fact that the Judge saw fit to impose such penalties is an indication of just how blameworthy Esso's behaviour was in his view.

1. Website at www.austlii.edu.au/au/cases/vic/VSC/2001/263.html.
2. The sentence, para 13 (see Appendix at p 61).

TABLE 1
Penalties

No	Count	Fine ($)
1	Inadequate hazard identification	250,000
	Further penalty	50,000
2	Failure to conduct an adequate risk assessment	150,000
3	Failure to maintain plant in a safe condition	200,000
	Further penalty	50,000
4	Inadequate procedures to deal with loss of warm oil circulation	100,000
5	Failure to adequately train employees to respond to a loss of warm oil	100,000
6	Failure to provide the means for ensuring that items of plant operated at a safe temperature	100,000
7	Failure to adequately train employees regarding the risks associated with abnormally cold plant	250,000
	Further penalty	50,000
8	Failure to adequately train supervisors regarding the risks associated with abnormally cold plant	150,000
	Further penalty	50,000
9	Failure to monitor conditions	200,000
10	Failure to ensure that the heat exchanger was protected from thermal shock	100,000
11	Failure to ensure the safety of persons other than employees	200,000
		2,000,000

Foreseeability of risk is one of the factors which courts take into account in determining culpability, and the foreseeability of this incident is evidently one reason why the Judge took such a serious view of the company's failures.[3] Moreover, the prosecution had argued that the gravity of the offence did not lie in the fact that two people had been killed. It was the potential for widespread death and injury in a major hazard facility such as a gas plant which made the failure to take appropriate safety measures so grievous. The Judge apparently accepted this view.

The purposes of punishment

Before embarking on a discussion of the reasoning which lay behind the sentences, it will be useful to review some of the purposes which punishment is supposed to serve. We shall consider here retribution, deterrence and rehabilitation.[4]

The historically fundamental purpose of punishment is retribution, that is, the idea that offenders deserve to be punished as a matter of principle, regardless of whether that punishment has any beneficial outcomes (such as deterrence or rehabilitation). The familiar concept of an eye for an eye and a tooth for a tooth is an expression of this idea. Although the Judge did not mention retribution explicitly, it is implicit in his discussion — a grievous offence deserves a severe punishment.[5]

A second purpose is deterrence. This can be of two types, general and specific. General deterrence refers to the deterrent effect of the punishment on others; specific deterrence refers to the effect on the person punished.

3. Thompson, W. *Understanding NSW occupational health and safety legislation*. Sydney: CCH Australia Limited, 2001, pp 54, 60.
4. There are others, for example, incapacitation, but these are not relevant here. For a fuller discussion see Bagaric, M, *Punishment and sentencing: a rational approach*, Sydney: Cavendish, 2001. Chapter 6 in Gunningham, N and Johnstone, R, *Regulating safety: systems and sanctions*, Oxford: Oxford University Press, 1999 provides a useful account of the purposes of punishment in the context of OHS offences.
5. Johnstone's analysis of sentencing in OHS cases in the Victorian County Court reveals that judges seldom refer explicitly to retribution but do refer to deterrence as one of the prime purposes of punishment. However, it is often clear from their comments about the need for penalties to reflect the "gravity" or "serious nature" of the offences that retribution is a major concern. Johnstone, R. *Safety, courts and crime: the legal construction of OHS offences in Victoria*. Sydney: Federation Press (in press).

General deterrence refers here to the possible effect of the Esso prosecution on other companies. The Judge at one point said that "punishment and general deterrence are of major significance in this case", indicating that one of the purposes of sentencing would be to send a message to other companies about what they can expect if they offend in similar ways.[6] But he did not elaborate on this and the goal of general deterrence remained implicit in his remarks. Whether the fine had any real deterrent effect on other companies is another matter. The disaster itself and the report of the Royal Commission had already caused many companies to pay much closer attention to these issues. It is unlikely that the prosecution and fine had any *additional* effect.

Moreover, for companies like Esso the potential costs of disaster are huge, and many times more than any realistically conceivable criminal fine. If these costs are not enough to persuade such a company to take the appropriate action, the threat of a fine which might be imposed in the wake of a disaster is unlikely to provide the necessary motivation.[7]

However, the Esso penalty may well promote general deterrence in another way. A likely consequence of the Esso fine is that penalties imposed in Victoria for OHS offences will rise as lower courts follow suit. This is likely to raise the profile of these prosecutions and hence enhance their general deterrent effect.

Moreover, the imposition of a fine of unprecedented magnitude has a moral educative effect on employers generally. It reinforces the legitimacy of OHS legislation and the values which underlie it. It is often observed that "the effective functioning of the criminal law is not based on fear, but on the legitimacy and acceptance".[8] In short, by enhancing the legitimacy of OHS law a large fine can have a preventive effect on the wider society that goes beyond general deterrence.

Moving to specific deterrence, this was certainly an explicit concern of the Judge, as will be discussed below. But it is worth noting here that in all the circumstances the specific deterrent effect of a $2m fine on Esso must be negligible. The explosion at Longford cost Esso $200m in lost sales and $100m to restore peak gas flow. Esso was also initially sued for over $1b in damages. These costs completely dwarf any conceivable fine. The theory of specific deterrence is that the penalty should ensure that the offender does not offend again. The fact is that the costs which Esso has already incurred or is threatened with as a result of the civil case have already had this effect. It has embarked on a $350m program to reduce site risks, to be completed by 2003. A fine cannot be expected to have any effect over and above the other costs which Esso has already incurred as a result of the incident.

There is another way, however, in which fines can be expected to have a specific deterrent effect. Inspectors who discover unsafe workplaces are entitled to launch prosecutions even in the absence of any major incident.[9] Such prosecutions are likely to result in real safety improvements.[10]

A final purpose of sentencing is rehabilitation. Human offenders can be sentenced to take part in various therapeutic activities which have some rehabilitative potential. Even corporations can be given rehabilitative sentences, for example, a directive to improve accountability mechanisms within the corporation.[11] A fine can hardly be said to have any rehabilitative value.

6. The sentence, para 39 (see Appendix at p 65). The context suggests that by "punishment" he meant "retribution".
7. See Hopkins, A, *Managing major hazards: the lessons of the Moura Mine Disaster*, Sydney: Allen & Unwin, 1999, ch 10 for a more extensive discussion of this point.
8. Bagaric, op cit, p 149.
9. Gunningham and Johnstone, op cit, p 185, describe these as pure risk prosecutions.
10. There is good empirical evidence from the US that fines for safety violations in the absence of any injury are effective in reducing injury rates. For a summary of the evidence, see Hopkins, A. *Making safety work: getting management commitment to OHS*. Sydney: Allen & Unwin, 1995.
11. Fisse, B and Braithwaite, J, *Corporations, crime and responsibility,* Cambridge: Cambridge University Press, 1993; Gunningham and Johnstone, op cit, ch 7.

However, rehabilitation can enter into the determination of the fine in the following way. In the case of a corporate offender such as Esso, if action has already been taken to reduce the likelihood of a further offence, then it can be inferred that some degree of rehabilitation has already occurred. A sentencer may decide that this justifies a lower fine. The Judge in the Esso case did mention rehabilitation as a consideration in sentencing but he did not develop the point.[12]

Corporate remorse

Because retribution is pre-eminent among the goals of punishment, the main factor influencing the magnitude of any sentence is the seriousness of the crime.[13] There are, however, other factors which can mitigate or reduce the sentence, one of these being whether the offender shows remorse or contrition. This can affect a sentencing decision in two ways.[14] First, it may reduce the gravity of the offence in the eyes of a judge, so that purposes of retribution may be served by a lesser sentence. Second, if the offender is truly remorseful it means that some measure of rehabilitation has occurred, that further offences are less likely, and that the need for specific deterrence is therefore lessened.

Human beings are capable of remorse, but it is a moot point as to whether corporations are capable of this emotion. There is a famous legal saying that a corporation "has no body to be kicked or soul to be damned", the implication of which is that remorse is not something which could be expected of a corporation.[15] Nevertheless, it was a corporation which was in the dock here and the Judge had therefore to give some consideration to the concept of corporate remorse and whether Esso's penalty deserved to be mitigated on this ground.

The Judge noted that Esso's senior officers had expressed personal remorse which, he had no doubt, was genuine.[16] But corporate remorse was another matter and there were three matters which led him to conclude that Esso as a company was not remorseful.

The first was what he called Esso's "litigious treatment of its employees". The only example he provided of this was Esso's attempt at the Royal Commission to blame one of the operators for the explosion. But this was a matter about which he was greatly concerned and he went to considerable lengths to stress that Esso's employees were loyal and brave men who were in no way to blame for the incident.

Second was Esso's conduct at the trial, described at length in Chapter 2. According to the Judge, "the defence advanced was one of obfuscation — designed not to clarify but to obscure".[17] This was hardly indicative of corporate remorse.

"The third matter, and the really significant one", said the Judge, "is the lamentable failure of Esso to accept responsibility for these tragic events".[18] During his final submission, counsel for Esso made the following statement: "Your Honour, Esso does wish to once again repeat its most profound regret for the loss of life and injuries that were caused to its employees by the explosions."

This statement does not admit any contribution by Esso, prompting the Judge to ask, "Does Esso accept responsibility for the fatal and injurious events?"

12. The sentence, para 40 (see Appendix at p 65).
13. This is the principle of proportionality. Bagaric writes: "The principle of proportionality in sentencing is a splendidly simple and appealing notion. In its crudest and most persuasive form it is the view that the punishment should equal the crime", and "Proportionality is most naturally associated with the retributivist account of punishment". Bagaric, op cit, pp 163, 182.
14. Hall, G. *Sentencing guide*. Wellington: Butterworths, 1994, p B221.
15. This saying was quoted at one point by the Judge, Ruling no 8, 20 March 2001, para 8.
16. The sentence, para 42 (see Appendix at p 65).
17. The sentence, para 45 (see Appendix at p 65).
18. The sentence, para 46 (see Appendix at p 66).

Counsel replied, "I can't answer that question in any blanket fashion".

The Judge was clearly concerned about this because a little later he asked again, "If Esso is the good corporate citizen you have stated it to be, why does it not accept responsibility...?", to which counsel again replied, "I can't answer that question".

The Judge explained his interest in this point as follows. "Normally in sentencing, a failure to accept responsibility sounds in the matter of specific deterrence", that is, the failure to take responsibility increases the risk of reoffending and necessitates a higher penalty to deter further offences. He went on: "This is not so here, because Esso has demonstrated by its subsequent operational reforms that it has acted responsibly to remedy past deficiency."

He then appeared to backtrack:

> "But its lack of acceptance of responsibility has a relevance in my not accepting that Esso's expression of remorse is practical and operational. Esso's failure to accept responsibility for these tragic events is a serious deficiency."

On one reading, this passage appears to say that despite Esso's risk reduction program since the explosion, its failure to take responsibility raised doubts about how effective this program was likely to be. It would follow from this that there remained a need to deter Esso from further offending.

But it may be that the Judge was reverting to the idea that an expression of remorse and a willingness to take responsibility lessen the gravity of an offence and therefore the need for retribution. The fact that Esso failed to accept responsibility meant that the gravity of the offence remained undiminished; there were no grounds here for any reduction in penalty.

The concept of corporate remorse, always strained, fails at this point. As was demonstrated in the previous chapter, Esso had a strong financial reason not to admit that its actions or inactions had caused the rupture and fire. To accept responsibility was to accept that it had caused the disaster and so it was rational corporate behaviour to deny responsibility, in the interest of the ultimate shareholding owners. Corporations are abstract entities without emotions. To expect a corporation to express remorse and to infer a lack of remorse from its failure to accept responsibility involves excessive anthropomorphism.

The plea

Another factor influencing sentencing is a plea of guilty.[19] Defendants are entitled to plead not guilty and should not be punished for so doing. At the same time, an offender who pleads guilty can expect a reduced sentence. Moreover, the earlier in the process the guilty plea, the greater the discount. Therefore, offenders who plead not guilty do so at their peril.

There are various reasons why a guilty plea is seen as deserving of a lesser sentence. One is that it may indicate remorse, which makes some reduction in sentence appropriate (for the reasons discussed above). A second is that it reduces the demands being made on the court system and is therefore to be encouraged — with a 10 to 25% reduction in penalty. Third, it shows consideration for vulnerable witnesses who are spared the pain of giving evidence when there is a guilty plea.

Esso did not plead guilty and so was not entitled to any reduction in sentence on these grounds.

Prior convictions

Another factor which influences sentencing is the presence or absence of prior convictions. The existence of prior convictions should not increase the penalty beyond what is warranted by the seriousness of the crime as that would be to punish the offender twice for the same offence.[20]

19. See the discussion in Thompson, op cit, pp 56-57.

20. Bagaric, op cit, p 230.

But prior convictions can override other mitigating factors to ensure no reduction in penalty.[21] Moreover, sentencers may take the view that the failure of previous punishment to deter indicates the need for a heavier penalty.[22]

In the case of corporate offenders, there is an interesting question about what should count as a previous conviction. If a company has previously been convicted of an offence under the *Trade Practices Act 1974* (for example, misleading advertising), should this count as a prior conviction in a prosecution under an OHS Act?

The Victorian OHS Act effectively answers this question by making explicit provision for further penalties in the event that a company has previous convictions under the OHS Act.[23] The Judge stressed that such a penalty was not a second penalty for the same offence; that would offend a fundamental principle of justice. Rather, it was to be understood as a contribution to a single penalty to be imposed for the current offence: "The further penalty marks the seriousness of present offences in the context of an offender who has previously offended."[24] In other words, the fact that the offender has been placed on notice by a previous similar offence, but has not taken effective action to prevent a recurrence, makes the present offence more serious than it would otherwise be. The further penalty is really a way of augmenting the initial penalty in order to reflect this greater seriousness.[25]

The Judge elaborated the criteria to be used in applying section 53 of the OHS Act as follows:

> "Circumstances justifying the operation of sec 53 are the nature and number of prior convictions, their proximity or remoteness in time to the present offence, their relevance, the character otherwise of the offender, and *whether the combination of prior convictions and present conviction demonstrates systemic failure by the offender or a longitudinal, general or flagrant failure to fulfil the lawful obligations of safety in employment.*"[26] (italics added)

The Judge's previous comments indicate that the italicised section of this statement is the real issue in determining the need for a penalty under this section.

Esso had three prior convictions which were brought to the attention of the Court. The first was in 1991, when Esso was the occupier of a service station site at Geelong. An LPG tank was being installed by an unrelated company and an employee of that company died as a consequence of entering the tank when it contained nitrogen. Both companies were prosecuted and Esso was convicted and fined $5,000 for failing to provide training and supervision to an Esso employee who supervised the deceased worker.

It might be argued that Esso was no longer in the service station business and that these events were unrelated to Longford. The Judge was initially uncertain as to whether this prior conviction was relevant, but he eventually decided that it was because it involved a training and supervision failure as had the Longford incident.

The other two convictions, in 1993, arose out of a fire on a Bass Strait production platform. Esso argued that these convictions were unrelated to the Longford events — they had occurred in a production plant, while Longford was a processing plant. But the Judge decided that these convictions

21. Bagaric, op cit, p 236.
22. Ibid, p 245.
23. Section 53.
24. The sentence, para 35 (see Appendix at p 64).
25. Johnstone, op cit, provides a useful analysis on earlier cases on this point.
26. The sentence, para 37 (see Appendix at p 64).

concerned failure to identify hazards and failure to provide safe plant, both matters on which Esso had been convicted at the trial.[27]

As a result, a further penalty was imposed for each of the three counts at the trial which mirrored these previous offences. But the Judge noted that "given Esso's otherwise very good safety record, I consider that the further penalties imposed should be moderate".[27] Accordingly, each of these further penalties was $50,000 or 20% of the maximum.

Totality

A further principle of sentencing comes into play when multiple charges arise out of a single incident or episode. In these circumstances there is a danger that, if the courts impose a sentence on each count independently and add these up arithmetically, the total sentence may be excessive. The principle of totality is that sentencers must have regard to the overall sentence when fixing sentences on each count, to ensure that the aggregate sentence matches in some way the aggregate of offences. As the Judge expressed it at one point, "sentencing ultimately is a holistic, not segmented, process".[28]

The Judge had determined at the outset of the trial that the counts did not involve duplication. But he recognised that there was a degree of overlap — the word he used was "interface". Accordingly, he moderated the penalties "to ensure there is no double punishment and no double counting". In so doing, he said, he would be "complying with the principle of totality".[29] Specifically, the Judge grouped the counts into categories, each category containing a group of overlapping offences. One count in each category was sentenced on its merits and subsequent counts in the category attracted reduced sentences. It should be noted that the Judge's application of the principle of totality compensates for the potential unfairness of charging Esso on multiple counts which was discussed in Chapter 2.

The Judge imposed the maximum possible penalty on two counts and it is worth commenting on the significance of these offences. The first count, perhaps not surprisingly, was the failure to conduct a HAZOP at Gas Plant 1. This was fundamental in the sense that many of the other failures were consequences of this initial failure. Safety starts with hazard identification and this had not been done.

The second offence which drew the maximum penalty was less predictable. It was the failure of Esso to train its employees about the risks they were subject to. Notice that this was not the failure to train employees in the procedures to be followed in certain circumstances, but the failure to ensure that employees were aware of the dangers they faced. The Court had been presented with a powerful image: men gathered near the heat exchanger which was about to explode, oblivious to the danger they were in, puzzling about how to get production going again and worried about the effect of leaks on Esso's obligations under environmental law. It was an image which weighed heavily with the Judge. As he said:

> "Those employees and their loved ones had every right to be properly trained about the risks Esso sent them to face. Esso totally failed in that most fundamental of matters."[30]

27. The sentence, para 38 (see Appendix at p 64).
28. The sentence, para 36 (see Appendix at p 64).
29. The sentence, para 49 (see Appendix at p 66). The principle has recently been restated in *NSW Dept of Education and Training v Keenan* [2001] NSWIRComm 106, discussed in Court Cases, *J Occup Health Safety — Aust NZ* 2001, 17(5): 446.
30. The sentence, para 50 (see Appendix at p 66).

Conclusion

The sentence and the remarks made by the Judge during sentencing revealed just how seriously he viewed the offences. Two of the offences attracted the maximum penalty, three attracted 80% of the maximum. Other fines were less so as to avoid double counting — not because the Judge viewed them less seriously.[31] It seems likely that had higher fines been available under the Act, the fines actually imposed would have been even greater.

Moreover, Esso's decision to plead not guilty, its conduct at the trial and its refusal to accept responsibility led the Judge to conclude that the company had shown no remorse, and the absence of corporate remorse weighed heavily in his decision not to mitigate the penalties in any way.

The sentence will no doubt be studied by lower courts in Victoria and possibly in other States and will be used as a guide in future sentencing decisions.

31. The sentence, para 50 (see Appendix at p 66).

CHAPTER 5

Learning from Longford

Following conviction, counsel for Esso addressed the Court on the issue of sentence. He devoted much of this submission to Esso's safety record and the risk-reducing activities carried out since the explosion. Although he still refused to admit Esso's guilt, he acknowledged the jury's findings, and asked that these findings be placed "in the context of Esso's commitment to safety generally, its safety record and indeed, its good corporate character".

These matters were potentially relevant to the sentence in various ways. Esso's behaviour *prior to* the explosion was relevant to the question of whether, given the existence of prior convictions, the offences warranted a further penalty. Esso managed to persuade the Judge that, apart from the Longford incident, it had a very good safety record.[1] The implication was that the Longford incident was to some extent an isolated event and not part of an extensive pattern of safety failures. On this basis the Judge chose to impose only a moderate additional penalty.

Esso's efforts to improve safety *following* the explosion were relevant in another way. They potentially demonstrated that Esso was remorseful, thus reducing the need for punishment. Unfortunately for Esso, in the mind of the Judge, its refusal to accept responsibility undermined the value of its safety improvements, which were apparently given very little weight in his sentencing decision.

Although ultimately to little avail, Esso's strategy was to *showcase* its safety achievements — to put them on display. For the outside observer this provides an invaluable opportunity to assess the extent to which Esso has learnt the lessons of Longford.[2]

As noted above, Esso's safety showcase persuaded the Judge of Esso's "otherwise very good safety record". However, several of the features he mentioned in this connection warrant a more sceptical response, in particular Esso's lost-time injury record, its safety management system, and its safety awards. None of the comments which follow should be read as a criticism of the Judge; he was not an expert in safety management and he was not in a position to question what he was told.

1. The sentence, para 38 (see Appendix at p 64).
2. I have not been able to gain access to Esso to examine first-hand how the company has responded to the explosion.

Safety performance monitoring

As was discussed at length in *Lessons from Longford*, lost-time injury rates (LTIs) are a dubious measure of safety.[3] They are influenced primarily by recording practices (for example, whose injuries are counted?), and by injury management strategies (for example, keeping injured workers at work). Moreover, LTIs are normally high frequency, relatively low consequence matters. Major hazard events such as explosions are rare, and do not generate LTIs for many years at a time. Hence, a focus on LTIs tends to divert attention away from the management of major hazards. This is exactly what happened at Longford. The Longford incident was of course an isolated incident, but it was part of a pattern of management failure which was obscured by the focus on LTIs.

Esso proudly told the Court that, during the period of restoration following the explosion, employees worked frenetically for 1.7 million work hours without an LTI. On the face of it, Esso is continuing to manage well the hazards which generate routine injuries, but its continued heavy (not exclusive[4]) reliance on this indicator in its submission suggests that it has yet to learn one of the important lessons about safety performance indicators for major hazard sites.

An excellent guidance note recently produced by the Victorian Major Hazards Division indicates how major hazard facilities should go about constructing more appropriate safety performance measures.[5] The logic is elegantly simple:

— Safety depends on hazard identification, risk assessment and risk control.

— Risk control means adopting control measures which are linked to specific hazards.

— Many major incident events occur because the controls which were supposed to be in place were not (for example, Piper Alpha and the failure of the permit to work system).[6]

— Ongoing safety depends on regular assurance that these controls are working as intended.

— Facilities must therefore adopt performance indicators which measure whether, or to what extent, controls are operating as intended.

— Finally, there must be measures of how well the management system is monitoring and maintaining these controls.

Table 2 presents examples taken from the guidance note. The importance of these indicators is that they measure how well the safety management system is performing with respect to *major hazards*, something which an LTI rate is incapable of doing.

It should be noted that the Victorian Regulations now require major hazard facilities to develop performance indicators of this type, and Esso has presumably complied.[7] However, no mention of this is made in its presentation on safety to the Court.

The indicators discussed in Table 2 are limited in some respects. They do not provide an overall or summary measure of how well major hazards are being managed. Moreover, because they are so site-specific they do not readily allow comparisons to be made between sites and they certainly cannot be used to make comparisons between industries. These are not defects, merely limitations — the utility of these indicators is limited to the purpose for which they are

3. Hopkins, A. *Lessons from Longford: the Esso gas plant explosion*. Sydney: CCH Australia Limited, 2000.

4. Another measure mentioned was the number of mishap-free flying hours for helicopters to platforms, although counsel for Esso did not place much value on this measure.

5. Victorian WorkCover Authority, Major Hazards Division. *Control measures and performance indicators under the Occupational Health and Safety (Major Hazard Facilities) Regulations* (MHD GN-10), September 2001. This is an outstanding document in an outstanding series. It deserves to be studied closely. Website at www.workcover.vic.gov.au/vwa/home.nsf/pages/so_majhaz_guidance.

6. Appleton, B. Piper Alpha. In Kletz, T. *Learning from accidents*. Oxford: Gulf, 2001.

7. Schedule 2, section 7.3.

TABLE 2
Examples of performance indicators for control measures

Examples of performance indicators for control measures
Management system compliance levels as shown by audit
Test frequency/interval for safety-critical equipment
Average skill level of the operations shift personnel
Compliance level with operating procedures as shown by monitoring
Number of failures in specific safety devices
Speed of response of safety device
Capacity (for example, flow rate, leak tightness) of safety device
Number of hazardous material releases per year
Inventories of different hazardous materials
Number of times staffing levels fall below target minimum numbers
Number of times pressure, temperature, etc exceed pre-alarm levels
Measured mechanical integrity (for example, amount of corrosion)
Detection and response time for accidental material releases
Number of occasions unregistered ignition sources are found on site
Sensitivity levels and response times for process alarms
Number of layers of protection for specified hazards
Deterioration limits on performance standards for any measures
Number of allowable simultaneous activities in process areas
Isolation and blowdown times for process units
Compliance level with segregation rules for dangerous goods stores
Compliance levels with manufacturer's or design standards
Vibration levels in rotating equipment (for example, compressors)

intended, that is, monitoring the performance of control measures.

Although not themselves summary indicators, they can provide the basis for a summary indicator. One of the first companies to provide a safety case under the Victorian *Occupational Health and Safety (Major Hazard Facilities) Regulations 2000* (MH Regulations) was Australian Vinyls. It defined a "critical event" as:

> "... any breach, failure or loss of a critical control measure or failure to meet a monitoring schedule for a critical control measure."[8]

Counting critical events provides a single, summary measure and the company declares that its goal is to drive this measure down to zero. Driving the critical event rate to zero is potentially far more relevant to the control of major hazards than driving the LTI rate to zero.

Australian Vinyls acknowledges that it previously used injury rate as the primary indicator of safety performance, and it attributes its new critical event measure to the learning which has come from the Longford incident.

It is useful to be able to make comparisons between plants or firms in a group since this stimulates a healthy competition to improve safety performance. However, measures of critical events at different plants are unlikely to be sufficiently similar in what they count to facilitate direct comparisons. One way around this is to ask all sites/firms in the "competition" to report not on the absolute number of events but whether that number has increased or decreased in the reporting period. To guard against spurious changes, for example, a decrease due to a change in the reporting procedure, "competitors"

8. Australian Vinyls Laverton Plant, Exemplar Safety Case, 30 September 2000, p 14, A70.

reporting a decrease should be asked to explain how they can be certain that the decrease is non-spurious. If the number of critical events in a reporting period is so low that increases and decreases are no longer statistically significant, the measure needs to be re-evaluated and additional events included. In this way, current competitions to drive down LTI rates can be replaced or at least augmented with competitions to drive down critical event rates.

Safety management systems

Another feature of Esso's safety showcase which the Judge acknowledged was its safety management system (SMS), known as OIMS (Operations Integrity Management System).[9] However, the Royal Commission was extremely critical of OIMS:

> "OIMS, together with all the supporting manuals, comprised a complex management system. It was repetitive, circular, and contained unnecessary cross-referencing. Much of its language was impenetrable. These characteristics made the system difficult to comprehend by management and by operations personnel.
>
> The Commission gained the distinct impression that there was a tendency for the administration of OIMS to take on a life of its own, divorced from operations in the field. Indeed it seemed that in some respects, concentration upon the development and maintenance of the system diverted attention from what was actually happening in the practical functioning of the plants at Longford."[10]

Although the trial Judge did not pick up on this problem in his sentencing remarks, counsel for the prosecution certainly did: "What this case highlights is that one cannot discharge one's duty by creating a monumental paper structure and then not implementing it."

Moreover, it cannot be forgotten that OIMS was the management system in place which allowed Esso to go on year after year without correcting the deficiencies which became so apparent at the Royal Commission and at the trial, and in particular without carrying out the HAZOP which Exxon had recommended.

Major hazard guidance note 12, which deals with SMS, has a useful classification of the way that SMS can fail. Two are particularly relevant here:[11]

> "The 'virtual' SMS is one containing the right management elements and addressing the correct control measures, but which does not reflect how these control measures are managed in practice.
>
> The 'misguided' SMS is one containing the right management elements but which manages the wrong control measures, ie those not critical to major incidents."

A virtual SMS, then, is a paper system which fails at the implementation stage, while a system which focuses on LTIs as its primary measure of safety is "misguided", from the point of view of controlling major hazards. One would have to say that, based on the evidence provided to the Royal Commission, Esso's SMS was in some respects both virtual and misguided.

Safety awards

The trial Judge also gave credit to Esso for its safety awards, among them a five-star rating award from the National Safety Council of Australia for SMS excellence. Again, there is reason for some scepticism about the significance of such awards. Some years ago a study was made of South African mines which used

9. The sentence, para 29 (see Appendix at p 63).

10. Dawson, D and Brooks, B. *The Esso Longford gas plant accident: report of the Longford Royal Commission.* Melbourne: Victorian Government Printer, June 1999, para 13.39-40.

11. Victorian WorkCover Authority, Major Hazards Division. *Safety management systems under the Occupational Health and Safety (Major Hazard Facilities) Regulations* (MHD GN-12), September 2001, p 7. Website at www.workcover.vic.gov.au/vwa/home.nsf/pages/so_majhaz_guidance.

the five-star International Safety Rating System. The study found no correlation between the number of stars a mine had received and its fatality rate or reportable injury rate.[12] This is an unnerving finding because it suggests that an award-winning SMS does not guarantee superior management of major hazards.

Other Esso safety awards mentioned approvingly by the trial Judge were based on the company's remarkable LTI rate. However, as noted earlier, such awards provide no indication of how well major hazards are being managed.

The material which Esso presented to the Court as evidence of its superior safety apparently persuaded the Judge that the Longford incident did not indicate a *systemic* failure and that Esso therefore did not merit a substantial additional penalty. But the preceding discussion has demonstrated that the evidence warranted a somewhat more sceptical response. Rather than demonstrating the absence of systemic failure, Esso's record masked it.

Safety improvements since the explosion

While the Judge apparently discounted the safety improvements which have occurred since the explosion (on the grounds that Esso had not accepted responsibility), the material presented to the Court about what Esso has done since the explosion suggests that the company has learnt some of the lessons very well.

A new focus on process upsets

The Royal Commission had criticised Esso for focusing on "slips, trips and falls" — minor personal injuries — at the expense of "process upsets", that is, situations where the process is running outside normal temperature, pressure and flow limits. Process upsets, if not managed properly, can lead ultimately to disaster. Esso's inappropriate emphasis in this respect was due to its obsession with the LTI rate as its primary safety indicator.

The company now has a new focus on process upsets and has trained its operators in how to respond to them. In particular, it runs "table top" process upset drills to enhance operator ability to deal with upsets.

Critical procedures

The company has also decided to classify procedures as either normal or critical. All tasks are being analysed to identify which procedures are critical. Manuals will reflect this distinction and staff attention will be focused on critical procedures.

Training

All staff have undergone additional "safety and integrity" training. Integrity refers to the integrity of the plant, and the juxtaposition of these two is a recognition that plant failure can have serious safety implications. In particular, there has been training with respect to cold temperature operations and the dangers of cold temperature.

Critical operating parameters and alarms

One of the problems revealed during the Royal Commission was that control room operators had to deal with hundreds, sometimes thousands, of process upset alarms each day. Many of these were no more than nuisance alarms, in the sense that they represented temporary deviations and the process could be expected to return to normality without operator intervention. It was appropriate simply to cancel these alarms. The problem was that no distinction was made between critical alarms and nuisance alarms, with the result that significant, indeed critical, alarms were often ignored or cancelled without proper consideration.[13]

12. Eisner, H and Leger, J. International safety rating system in South African mining. *J of Occupational Accidents* 1988, 10: 141-160.

13. This problem of alarm overload is not unique to Esso. I recently visited a gas processing plant where alarm overload was so acute that one operator suffered a repetition injury from cancelling alarms. It is noteworthy that the problem of alarm overload came to light because of this injury and not before, indicating that at this plant, too, LTIs were taken more seriously than process upsets.

Esso has since expanded its use of critical operating parameters at Longford to more closely define the safe operating envelope of the plant. A total of 111 such parameters have been identified for the whole site. When critical operating parameters are exceeded, an alarm registers and there is now a system to ensure that operators fully investigate any such alarms and take appropriate action. Table top drills are designed to ensure that operators are familiar with these procedures.[14]

All of this is required under the new MH Regulations and so Esso was in a sense making a virtue out of necessity.[15] But this is not to belittle these changes; they merited a more positive response from the Court than they in fact received.

Supervision

Esso has recognised that the above changes involve higher levels of activity and has created one additional supervisory position. In an era of downsizing and cost-cutting this is a noteworthy development, although it seems unlikely that one post can make up for the previous reduction in supervision criticised by the Royal Commission.[16]

Hazard identification

All existing plant will be HAZOPed every five years. This policy substantially increases Esso's hazard identification activity and deals with what can reasonably be described as one of the root causes of the explosion.

Remote monitoring

Esso has replaced the outdated monitoring system (including the paper and ink recording charts) with a fully computerised system. This system now delivers to head office in Melbourne a great deal of information which enables engineers there to monitor process upsets and critical operating parameter excedences. They will therefore be able to keep the plant under far more effective surveillance than in the past.

However, Esso does not appear to have relocated engineers to the Longford site. Its failure to do so puts it at odds with the Royal Commission and with a recent report by the Institution of Engineers, Australia. The report, entitled "Have Australia's Major Hazard Facilities Learnt from the Longford Disaster?", included the following recommendation:[17]

> "Major hazard facilities must have access to sufficient engineering, operating and maintenance skills on site and at all times coupled with regular comprehensive surveillance of operating practices and properly kept records."

An anonymous reporting system

One of the changes which have been made since the Longford explosion is the introduction of an anonymous reporting system. This system apparently avoids identifying either the reporter or any person involved in the reported incident. According to counsel for Esso, there was a reluctance to use the existing incident and near-miss reporting system because "it might be seen as dobbing in your mates". The way to overcome this problem was to introduce a system in which no one was identified.

The new reporting system prompts various observations. First, it is widely agreed that the single most important way to promote safety is to encourage the reporting of incidents and near misses.

14. In one control room I walked into recently the first thing I saw was a critical alarm illuminated on the control panel. The alarm was being ignored because when certain parts of the plant were out of action this alarm normally appeared and operators "knew" that it did not need to be attended to. One hopes that such outcomes will be avoided by the procedures which Esso has put in place.

15. Schedule 4, B 1(1)(b) and (2).

16. Longford Royal Commission Report, op cit, p 198.

17. Published by the Institution of Engineers, Australia, October 2001, p 30.

Unless an organisation can create a climate which is favourable to reporting, it cannot claim to have a safety culture.[18] One of the prerequisites for an effective reporting system is that the organisation has a no-blame culture, that is, that people who report things which are detrimental to themselves or others are not blamed but, on the contrary, are commended. If no blame is attributed to either the reporter, or the person about whom the report is made, assuming there is such a person, then the problem of dobbing on your mates disappears. Esso's initiative can be read as an implicit recognition that a culture of blame operates within the organisation. Freely flowing communication about incidents can only be expected if Esso attacks the culture of blame directly and gives due recognition to reporters.

A second observation to be made about Esso's new scheme is that it presumes that the incidents it seeks to have reported concern the behaviour of individuals who may be embarrassed by these reports, and that reporters are unwilling to report for this reason. The Royal Commission produced a good deal of evidence that suggested that these presumptions are wide of the mark.

Vital information about safety-critical incidents was in fact recorded in operator log books. In particular, there were reports of cold temperatures which operators could not explain or correct, which in retrospect can be seen as precursors of what happened. Had they been investigated, the explosion at Longford could well have been avoided. The problem here is not that operators were unwilling to report for fear of dobbing in their mates. On the contrary, the operators were quite willing to report. The problem was that the system at the time did nothing with their reports.

Or again, take the cold temperature incident which occurred a month before the disaster and which was in many respects a dress rehearsal for it. Operators did not refrain from reporting this incident to avoid dobbing in their mates. They simply didn't think to report it because it was not the sort of thing they were encouraged to report.

It may well be true that some information is not reported because of reluctance to dob in mates, but a great deal of vital information fails to find its way into reporting systems because workers are not encouraged to report such matters or because the system fails to pick up and process the reports which are made. Anonymous reporting systems do nothing to overcome these problems.

An example of a well-focused incident reporting system is provided in the Australian Vinyls safety case. It specifies that all critical events, as defined above, be recorded in the incident reporting system and that all such events be reviewed with senior management involvement.[19] Other ways in which reporting systems can operate more effectively are discussed in *Lessons from Longford*.[20]

Conclusion

Esso placed its safety achievements on display during the sentencing phase of the trial. It did so in order to persuade the Judge that it warranted a lenient sentence. I have argued that the Judge was unduly impressed by Esso's past record and insufficiently impressed by Esso's efforts since the explosion.

Be that as it may, by showcasing safety in the way that it did, Esso offered outsiders an insight into the lessons which it has learnt from the Longford explosion, as well as the lessons which it has still to learn.

18. Reason, J. *Managing the risks of organisational accidents*. Aldershot: Ashgate, 1997.
19. Australian Vinyls, op cit, p 34.
20. Hopkins, op cit, ch 5.

CHAPTER 6

Corporate manslaughter

Despite the unprecedented fine, the sentencing of Esso was followed by renewed demands for a stronger criminal justice response to industrial fatalities. One union spokesperson noted that under the *Trade Practices Act 1974* companies can be fined up to $10m per count and declared that much higher penalties were needed in OHS legislation.[1] But he went further than this: "Esso's senior officers should have been faced with charges of manslaughter." According to another union official: "Esso's lack of corporate remorse shows the need for industrial manslaughter legislation to force companies to bear responsibility for their actions." Even the Judge in his sentencing remarks referred to limited penalties available under the *Occupational Health and Safety Act 1985*, hinting that he would have imposed higher fines had that been possible.

The Victorian Labor Party had for some time been concerned about what it saw as the inadequate penalties being imposed where workers lose their lives in circumstances of company negligence. One of its election promises when it came to power in 1999 had been to introduce industrial manslaughter legislation. At the time of Esso's sentencing in July 2001, the legislation had been drafted and was out for community consultation. Union officials were keen to see it enacted quickly. As one put it, the Esso "case should be used by the community as justification for the new legislation".

A Bill defining the offence of corporate manslaughter was presented to the Victorian Parliament towards the end of 2001 and was passed by the Lower House but defeated in the Upper House in June 2002.[2] The Labor Party has not abandoned this project, however, and the Bill may well be resurrected if and when the political circumstances allow. The ACT Labor Government has also indicated that it intends to introduce industrial manslaughter legislation, and similar legislation is contemplated in the UK.[3] It seems likely, then, that sooner or later we will see some form of industrial manslaughter law enacted in Australia.

This raises an intriguing question: had the proposed Victorian legislation been in force at the time of the Longford incident, is it possible that Esso might have been convicted of manslaughter? Attempting to answer this

1. The statements by union officials in this paragraph and the next are taken from *Inside OHS*, August 2001, no 16, pp 1-2.

2. Victorian Parliament. *Crimes (Workplace Deaths and Serious Injuries) Bill 2001.*

3. Wells, C. *Corporations and criminal responsibility* (2nd ed). Oxford: Clarendon, 2001, p 124.

question will provide a fresh perspective on the Esso incident and will provide us with an opportunity to consider the significance of the proposed legislation for large firms in general.

Existing law on manslaughter by corporations

In principle, it is possible in Australia to prosecute a corporation for manslaughter. However, no large company has ever been convicted here because of the difficulty of establishing that a corporation has the necessary *mens rea*, or criminal state of mind. The state of mind required for a manslaughter conviction is gross negligence, and the legal rule is that if some company officer, who is sufficiently senior to be regarded as the mind of the company, has exhibited gross negligence, then this state of mind can be attributed to the company and the company convicted of manslaughter. The problem is that it can rarely be said that the most senior company officials have displayed gross negligence in relation to the circumstances of a fatality. In large companies they are usually far removed from those circumstances. Thus, large companies are virtually immune from manslaughter charges. In contrast, where a company is little more than an incorporated individual or a partnership, a director may be personally responsible for the circumstances surrounding a fatality. If so, it is comparatively easy to establish the company's guilt.[4]

At least one celebrated attempt to prosecute a large company has come to grief because of the problem of corporate *mens rea*. Following the capsize of an English Channel ferry in 1987 (in which 188 people were drowned), an inquiry found that there was negligence on the part of individuals throughout the company: "… from top to bottom the body corporate was infected with the disease of sloppiness."[5] But a subsequent attempt in the UK to prosecute the owner, P&O, failed because no one individual, and in particular no individual at the top, could be said to have been grossly negligent in relation to the sinking.

Legal commentators have long been suggesting that the law of manslaughter needs to be changed to reflect the realities of corporate negligence.[6] Decision-making in large corporations is diffuse and courts need to be able to aggregate the negligence at various levels of the company. This was the real purpose of the Victorian Bill.

The Victorian Bill

The first point to note about the *Crimes (Workplace Deaths and Serious Injuries) Bill 2001* is that it did not seek to introduce the new offence into the OHS Act but rather into the Victorian *Crimes Act 1958*. This has considerable symbolic significance. Notwithstanding the discussion in Chapter 3, there is still a tendency in some quarters to regard OHS offences as less than criminal. Co-locating the offence of corporate manslaughter with a range of other heinous offences in the *Crimes Act* is a clear statement, if one was needed, that the offence is truly criminal in nature.

There is, however, a detrimental aspect to locating the offence in this way. As Johnstone noted:

> "The prosecution of those causing workplace fatalities under the general criminal law may be counter-productive, in that by focusing on a few particularly serious cases and singling them out for special treatment in the form of manslaughter prosecutions, an OHS enforcement agency risks undermining the 'normal' OHS offence under OHS legislation."[7]

Second, the Bill created two new corporate offences, that is, corporate manslaughter and negligently causing serious injury by a corporation. The two

4. One small company in Victoria has been convicted of manslaughter in this way — Johnstone, R. *Occupational health and safety law and policy*. Sydney: Lawbook Company, 1997, p 432.
5. Wells, C. *Corporations and criminal responsibility*. Oxford: Clarendon, 1993, p 47.
6. Fisse, B. *Howard's criminal law* (5th ed). Sydney: Lawbook Company, 1990.
7. Johnstone, op cit, p 431.

offences ran parallel in many respects and the present discussion will be restricted to corporate manslaughter.

Third, the Bill did seek to amend the OHS Act by increasing maximum penalties and, in particular, by raising the maximum fine for failure to maintain a safe workplace from $250,000 to $600,000.

The offence of corporate manslaughter

The offence of corporate manslaughter is laid out in a series of sections. First, section 13:

> A body corporate which by negligence kills an employee, or contract worker, is guilty of the indictable[8] offence of corporate manslaughter and liable to a fine not exceeding $5m.[9]

The nature of this offence is very different from an offence under the OHS Act. The OHS Act effectively makes negligence (failure to do what is practicable) an offence, regardless of whether that negligence results in a death or injury. Under corporate manslaughter legislation, negligence would be an offence only if it resulted in death. The question of whether the negligence caused the death is therefore a potential issue. In many cases causation will not be in dispute, but in the Esso case it was. For a manslaughter charge to succeed against Esso, the prosecution would need to establish causation, but since the jury did indeed find that Esso had caused the incident, this would not have been an impediment to a manslaughter conviction.[10]

A further observation about section 13 concerns the $5m penalty that it envisages. Although this is much greater than the maximum penalty per count under OHS legislation, it is not dramatically greater than the penalty which can be imposed when a company is charged with multiple counts, as Esso was. The Bill in fact raised the penalties available under the OHS Act by 240%, and if we apply this percentage increase to the total fine imposed on Esso, it comes to $4.8m. From this point of view, the new offence did not involve significantly higher penalties than those proposed under the OHS Act. Its significance, as suggested earlier, was symbolic. The stigma of a conviction for manslaughter is far greater than the stigma of conviction under the OHS Act. In so far as the criminal law has any deterrent effect on corporations, a manslaughter conviction can be expected to have a greater effect than an OHS conviction.[11]

The truly organisational nature of corporate manslaughter is recognised in section 14B(4):

> "In determining whether a body corporate is negligent, the conduct of the body corporate as a whole must be considered."

The problem of corporate *mens rea* which has frustrated previous manslaughter cases, as described above, is overcome in section 14B(5):

> "… the conduct of any number of employees, agents or senior officers of the body corporate may be aggregated …"

Section 14B(6) provides five examples of the sorts of things which courts may take as evidence of negligence. Two are worth noting because of their particular relevance to the Esso case:

> "(a) failure adequately to manage, control or supervise the conduct of one or more of its employees, agents or senior officers …
>
> (c) failure to provide adequate systems for conveying relevant information to relevant persons in the body corporate …"

8. Indictable means triable before a judge and jury.
9. This is not an exact quotation, being slightly abbreviated and simplified. In particular, the penalty in the provision is specified as 50,000 penalty units which equates to $5m.
10. See count 10, discussed in Chapter 2.
11. The question of whether criminal convictions in fact deter, either in the specific or general sense, is a complex one which is not addressed here. For some useful discussion see Fisse, B and Braithwaite, J. *Corporations, crime and responsibility*. Cambridge: Cambridge University Press, 1993.

Gross negligence

It was not envisaged that all negligence would render companies liable, only gross negligence. As the Minister put it in his second reading speech, "the degree of negligence required is gross negligence, that is, criminal negligence, rather than [civil] negligence".[12]

Section 14B(1) spells out what this means. It states that the conduct of a corporation will only be regarded as negligent, for the purposes of the Act, if it involves:

> "(a) such a great falling short of the standard of care that a reasonable body corporate would exercise in the circumstances; and
>
> (b) such a high risk of death or really serious injury — that the conduct merits criminal punishment for the offence."[13]

This is perhaps the single most important section of the Bill in terms of understanding whether or not a corporation might be potentially guilty of corporate manslaughter. Several observations are therefore in order.

First, there is a certain circularity about this section.[14] It purports to give guidance about the type of negligence which would be regarded as gross or criminal for the purposes of the legislation. It would be helpful for a jury to have some independent criteria of criminal negligence and to be able to say: because the negligence meets these criteria, we conclude that it is criminal negligence and therefore deserving of criminal punishment. Instead, the jury must first decide whether the behaviour is so far short of what is reasonable, and so risky, that it deserves criminal punishment. If so, it follows as a consequence that the behaviour can be described as gross or criminal. It is apparent, then, that the term "gross" is of no assistance to the jury in deciding whether the defendant corporation is guilty. As if in implicit recognition of this fact, the word "gross" appears nowhere in the Bill.

The real issue for a jury, then, is not whether the behaviour is grossly negligent but whether the risk of death is sufficiently high to be condemnable and whether the conduct is so far short of reasonable as to be culpable. I concentrate here on the second of these questions.

How can a jury decide whether conduct is so far short of the way a reasonable corporation would behave as to be culpable? Juries are presumed to consist of reasonable people. In the case of individual manslaughter, therefore, in order to decide what a reasonable person would do, the members of a jury can sometimes (not always) substitute themselves and come to conclusions about how far short of reasonable behaviour the conduct of the defendant is by examining their own responses, that is, by a process of introspection. However, in deciding what the reasonable corporation would have done, they can never rely on introspection in this way. For example, to refer to the Esso case, how far short of reasonable corporate behaviour was Esso's failure to conduct a HAZOP? To answer this question would require evidence. What is the industry practice in regard to HAZOPs? What would other companies do when faced with recommendations from parent companies about HAZOPs? This is very much the question which Kletz mused about from the UK perspective:

> "Was Longford a small plant in a distant country that fell below the company's usual standards or did it indicate a fall in standards in the company as a whole?"

If the answer is the former, then perhaps a jury would be in a position to find Esso guilty of corporate manslaughter. But if the answer is the latter, namely that Esso's behaviour was in line with Exxon's falling standards, then it would be harder to find Esso Australia guilty. The point is that these are questions which no jury could decide without evidence.

12. Victorian Parliament. *Hansard*, 22 November 2001, second reading of *Crimes (Workplace Deaths and Serious Injuries) Bill 2001*.

13. This is basically a restatement of the common law, see *Nydam v R* [1977] VR 430 at p 445.

14. And therefore in the common law.

It is, however, possible that juries might decide the matter using more *a priori* reasoning, that is, without evidence of industry standards. For example, it is possible that a jury might conclude that, regardless of what industry practice is, it is totally unreasonable to fail to carry out a HAZOP when asked by the parent company to do so.

In Esso's case there is one additional indicator of the way things might have gone. It will be remembered that the Judge was willing to conclude on the evidence that the "cause [of the incident] was grievous". The jury had essentially found that the cause was Esso's negligence (failure to do what was practicable) and the Judge was therefore saying, in effect, that Esso's negligence was grievous. It is surely a short step from "grievous" to "gross". There is a realistic possibility, therefore, that even on the evidence presented at the trial, a jury might have been willing to convict Esso of the offence of corporate manslaughter as defined in the Victorian Bill.

The Government's intention

Whether the authorities would ever have initiated a prosecution against Esso for manslaughter is another matter. The Minister noted in his second reading speech that the legislation might apply to building sites "where there may be a lack of safety equipment, a lack of training and a lack of supervision". This was clearly a reference to the spate of building industry fatalities in recent years in Victoria.

But he was also at pains to explain that the legislation would be invoked only in exceptional cases:

> "This Bill is designed to catch those rogue operators who think that they can get away with, or do not care whether they are, running an unsafe workplace... It is only [for] the most evasive or irresponsible employers."

These descriptions do not apply to Esso. Even the prosecutor at the trial conceded that "this is not a case of a rust bucket where no one cared for safety. On the contrary".

The prosecuting authority exercises its own discretion in deciding whether to prosecute, but if we assume that this discretion is exercised in a manner consistent with government intentions, it would seem unlikely, in the normal course of events, that Esso would have been charged with manslaughter. On the other hand, the course of events was not normal. The political circumstances surrounding the case called for a dramatic response from the government of the day — hence the Royal Commission.[15] Those same political circumstances led to the decision to run the matter in the Supreme Court, ensuring the highest profile OHS trial ever conducted in Australia. Had corporate manslaughter legislation been on the books there is no guarantee that it would not have been used in the Esso case.

Individual liability

Although the Bill was primarily aimed at creating corporate liability for manslaughter, section 14C also specifies individual liability. If, and only if, a corporation was found guilty of corporate manslaughter, a senior officer might also be convicted of an offence in some circumstances. The penalty was up to five years' imprisonment.

For a senior officer to be found guilty, the prosecution needed to show that:

— he or she was organisationally responsible for the relevant corporate conduct;

— the failure to carry out his or her responsibilities contributed materially to the outcome;

— the senior officer knew that as a result of his or her failure there was a substantial risk of the outcome; and

— having regard to the circumstances known to the senior officer, it was unjustifiable to allow the risk to exist.

15. See the discussion in Hopkins, A. *Lessons from Longford: the Esso gas plant explosion*. Sydney: CCH Australia Limited, 2000, pp 2-3.

It is hard to imagine any senior officers of a large corporation being caught by these provisions because it would be very difficult for the prosecution to show that they actually knew of the possible consequences of their failures. As the Minister explained:

> "The Bill requires that a senior officer must actually know that there is a substantial risk of death or serious injury to an employee. Like other serious criminal offences, it is important that the senior officer's liability be based on what the senior officer actually knows."

There was no suggestion at the Esso trial that any of its senior officers might have been liable in this way.

In fact, as a result of lobbying by business groups,[16] the offence was defined in such a way that senior officers were even less liable to be prosecuted under the Bill's provisions than they were to be prosecuted for manslaughter under the existing common law.[17] Given that no senior officer of a large company has ever been prosecuted for manslaughter in an industrial context in Australia, one would have to say that, on the whole, senior company officers had nothing to fear from the Bill.

This was certainly a disappointment to the unions. They had hoped that the proposed legislation would make senior executives more accountable than they are at present.

As one union leader said:

> "Monetary penalty is not enough and that's why the industrial manslaughter legislation is absolutely vital because it will take a few directors of companies to be pinged for industrial manslaughter before some of the boardrooms and the executives will take the issue seriously."[18]

The Victorian Bill did not achieve this objective.

A concluding note

There has been much talk in Victoria in recent years about the possibility of so-called "industrial manslaughter" legislation. The proponents of such legislation have always assumed that it would target individuals — company directors and senior executives. But the Victorian Bill targeted corporations, not individuals. In this respect it embodied suggestions which legal writers have been making for a number of years and it reflects the truly organisational nature of much corporate offending.[19]

Nevertheless, there are good arguments for targeting individual senior officers who fail to take safety seriously.[20] At present this is possible under the Victorian OHS Act only if the failure was deliberate or the result of "wilful neglect".[21] In NSW and Queensland, prosecution is possible under OHS legislation if senior officers fail to exercise "due diligence" in relation to safety.[22] This is easier for courts to establish and there have indeed been convictions under these provisions. Victorian law should be amended in this respect to bring it into alignment with NSW and Queensland.[23]

16. *The Age*, 23 August 2001.
17. The offence was one of recklessness. In an earlier draft it was not required that the senior officer actually "knew" — "ought to have known" was sufficient, and the offence was one of negligence.
18. *The Age*, 29 June 2001.
19. For example, Wells, 1993, op cit.
20. See Fisse and Braithwaite, op cit.
21. Section 52(1).
22. NSW *Occupational Health and Safety Act 2000*, section 26(1); Qld *Workplace Health and Safety Act 1995*, section 167(4).
23. An earlier draft of the Bill did contain this amendment. But, presumably as a result of intense lobbying, it was deleted from the final version of the Bill presented to the Victorian Parliament.

CHAPTER 7

Conclusion

Chapter 1 posed a series of questions which this study promised to address. Succeeding chapters have answered these questions, but sometimes only implicitly. The final chapter returns to these questions and provides explicit answers.

Why was the trial held in the Supreme Court?

The prosecution argued, and the Judge agreed, that the matter was of such public importance that it needed to be heard in the Supreme Court so that authoritative legal rulings and sentencing principles could be established to guide the lower courts.

Did the trial produce any new findings about the causes of the disaster?

The purpose of the trial was not to make findings about what happened — that was the central purpose of the Royal Commission — but to determine whether Esso was at fault. The very structure of a criminal trial militates against detailed or nuanced findings; in principle, it is a contest between parties to produce one simple finding — guilt or innocence. It did not, therefore, produce any new causal insights.

Moreover, the trial focused on only one part of the accident sequence, that is, what happened after the warm oil pumps failed. There was much to be learnt about incident prevention by examining earlier parts of the sequence, but Esso's culpability was greatest in relation to the subsequent events, so this was where the prosecution concentrated.

How did the confusion between "risk" and "hazard" affect the trial?

The trial was essentially about the failure of Esso's hazard management process. There are three fundamental steps in this process: hazard identification; risk assessment; and risk control. Esso carries out two different types of hazard identification, one called a HAZOP (hazard and operability study) and one called a periodic risk assessment. Calling the latter a risk assessment confused matters at the trial. The Judge understandably thought that a periodic risk assessment was something which followed hazard identification in the above sequence. The result was total confusion about just what Esso was charged with in count 2.

The problem stems from inconsistencies in the use of the word "risk" by risk management professionals. It is exemplified in AS/NZS 4360 which uses the word in two distinct ways — on the one hand, hazard, and on the other, severity and likelihood of an incident.[1] The confusion in the Esso trial was a direct manifestation of this terminological problem.

1. AS/NZS 4360-1999: *Risk management*. Sydney: Standards Australia, 1999.

What does "practicably preventable" mean?

Safety failures are offences only if they were practicably preventable. The trial provided a useful indication of what this means for a large company like Esso. At law there are four things to be considered:

1. the severity of the hazard or risk in question;
2. the state of knowledge which the employer had or ought to have had about that hazard or risk and about any ways of removing or mitigating that hazard or risk;
3. the availability and suitability of ways to remove or mitigate that hazard or risk; and
4. the cost of removing or mitigating that hazard or risk.

In a context such as Esso's gas plant at Longford, the first, third and fourth of these are not in question: the hazards are severe, there are ways to control the hazard, and the cost for a large company like Esso is not prohibitive. The only remaining question is the second, whether the employer ought to have foreseen the possibility of what happened. Practicability therefore boils down in this case to foreseeability. If the incident was foreseeable, it was practicably preventable. This will not always be the case, but it is clear that an employer who foresees a possible incident but does nothing about it on the grounds that it is not practicable to take preventive action is running a considerable legal risk.

Why was Esso charged on so many counts — 11 in all?

The prosecution appeared to "throw the book" at Esso by preferring 11 counts. One consequence of this was the imposition of a much larger fine than would otherwise have been possible. But that was not the purpose. The prosecution wished to highlight the multitude of failures which had led to the explosion. At law, every failure must be the subject of a separate count so that the accused knows exactly what is being alleged. Hence the multitude of counts. This led to concerns about whether some of the counts were so similar that Esso was in effect being charged with the same offence twice. The Judge ruled that this was not the case, but to ensure that there was no double punishment he imposed a reduced penalty on charges which appeared to overlap.

Are OHS offences crimes?

Offences under the OHS Act differ from normal criminal offences in that it is not necessary to show that the offender intended the act or knew it was wrong. They are strict or absolute liability offences. It is sometimes argued that for this reason they are less than truly criminal.

An alternative view is that crimes are offences to which criminal procedures apply, in particular the requirement that proof be beyond reasonable doubt. Offences under the OHS Act are criminal in this sense. Defendants are entitled to all the protections available in criminal trials and the Esso trial was a model of criminal procedure in this respect. The Judge was very clear that the offences with which Esso was charged were crimes and said so on numerous occasions.

How effective was Esso's defence strategy?

Esso's strategy was to offer no explanation at the trial for the incident. To have done so would have meant putting experts in the witness box and exposing them to cross-examination which might have elicited damaging admissions. Instead, Esso confined itself to trying to cast doubt on every aspect of the prosecution's explanation. A possible reason for this strategy is that it wished at the later civil hearing to contest the claim that its conduct was the cause of the incident, in order to escape liability for damages.

The strategy, however, was a failure. Not only did the jury effectively find that Esso's conduct was the cause of the explosion, but the defence antagonised the Judge, leading him to describe it as designed to obscure. This had a significant effect on the penalty.

Why did Esso not blame its control room operators at the trial, when it had blamed them at the Royal Commission?

Esso did its utmost at the trial to avoid blaming the operators. Most dramatically it declined to cross-examine the operator whom it had pointedly blamed at the Royal Commission. Blaming the operator at the Royal Commission had backfired badly in the court of public opinion and Esso could be sure that any suggestion at the criminal trial that it blamed the operators would inevitably antagonise the jury. Hence the about-face.

Why did the Judge say so emphatically: "What happened was no mere accident"?

Esso had suggested at the trial that the explosion was merely an accident. In the Judge's view it was the foreseeable outcome of a series of failures for which Esso was thoroughly responsible. The claim that it was an accident was a denial of responsibility and a refusal to acknowledge the causal process which had been well established in both the Royal Commission and the trial. Hence the Judge's insistence.

How culpable was Esso?

The Judge found Esso to be extremely culpable. His statement that the causes of the incident were "grievous, foreseeable and avoidable" could hardly have been stronger. This impression is reinforced by his use of the maximum possible fines on two counts and fines which were 80% of the maximum on three others.

Esso's culpability stemmed not only from the offences themselves but also from Esso's failure to demonstrate remorse. This weighed heavily in the Judge's sentencing decisions.

What meaning can be given to the concept of corporate remorse?

The presence or absence of remorse is an important determinant in sentencing. But remorse is a human emotion. How can a corporation show remorse? The Judge identified three ways:

1. by treating the employees concerned with respect and compassion;
2. by conceding more than Esso did at the trial (the logical extension of course is to plead guilty); and
3. by admitting responsibility (if not at the trial, then certainly during the sentencing hearing).

What kinds of prior corporate convictions count for the purposes of sentencing?

Under Victorian law, further penalties can be imposed in the light of prior OHS offences. The Judge imposed further penalties only where the prior offences were of the same kind: for example, a training failure received a further penalty because there was a previous training failure, and a failure to maintain safe plant was further penalised because of a similar previous conviction.

The Judge provided an important rationale: a similar prior conviction pointed to a systemic failure. It was not just a matter of imposing a higher sentence for a second offence; it was because the second offence demonstrated an ongoing and unrectified problem.

To what extent has Esso learnt the lessons of Longford?

Esso used the occasion of the trial to present the safety improvements that it had made since the explosion. On the basis of this presentation it is clear that there have been major improvements. Perhaps the main lesson of the explosion was the need for a much greater focus on process upsets. Esso has developed such a focus. It has defined critical operating parameters, reformed its alarm system accordingly, introduced procedures to be followed when critical alarms occur, and trained its operators in these procedures. It has also carried out numerous hazard identification exercises, and it has substantially improved its process monitoring capabilities.

However, Esso's primary measure of safety performance remains its LTI rate (which reveals nothing about how major hazards are being managed). Australian Vinyls has provided a model in this respect — a critical events measure — which can be used to monitor performance with respect to the management of major hazards, but Esso provided no evidence to the Court that it had adopted any such measure.

Finally, one of the lessons of Longford drawn by the Royal Commission was the need for on-site engineers. Based on its presentation to the Court, Esso does not appear to have learnt this lesson.

Would Esso have been at risk of conviction under the corporate manslaughter legislation which was proposed for Victoria?

The proposed corporate manslaughter legislation required a jury to find that a company has exhibited gross negligence in order to be guilty of manslaughter. It is not impossible that a jury would have been willing to find Esso's negligence gross.

Many of the proponents of industrial manslaughter legislation hoped that it would be directed at individuals and make it possible to send company directors to gaol in some cases. But because of the way the proposed legislation was drafted, directors of large companies had nothing to fear from it. Individual liability of senior officers would be better achieved by reforming the Victorian OHS Act on this point to bring it into line with other States.

Concluding comment

The Esso trial is a landmark. It has set new standards for the conduct of trials and it has enshrined hazard management as the duty of all employers. It has raised the level of fines in this country to new heights and provided a powerful statement of the culpability of firms which fail to provide a safe workplace. It adds a moral dimension to the official response to the Longford disaster and, in so doing, it transcends the Royal Commission.

APPENDIX

Director of Public Prosecutions v Esso Australia Pty Ltd

Sentence

JUDGE:	Cummins J
WHERE HELD:	Melbourne
DATE OF PLEA:	29 June 2001
DATE OF SENTENCE:	30 July 2001
CASE MAY BE CITED AS:	DPP v Esso Australia Pty Ltd
MEDIUM NEUTRAL CITATION:	[2001] VSC 263 Revised 6 August 2001

In the Supreme Court of Victoria at Melbourne Criminal Division No 1484 of 2000

CRIMINAL LAW — sentencing — workplace safety — convictions under sec 21, 22 and 47 *Occupational Health and Safety Act 1985* — gas processing plant at Longford, Victoria — major hazard facility — catastrophic failure — death and personal injury in workplace — risk to non-employed persons — unsafe workplace and systems of work — lack of safety training — further penalty under sec 53(a)(i) — considerations applicable in sentencing — *Occupational Health and Safety Act 1985,* sec 4, 6, 21, 22, 47 and 53.

Appearances: Counsel for the prosecution, R Richter QC, with N Clelland; solicitors from the Office of Public Prosections. Counsel for the defence, M Titshall QC, with M Hennessy; solicitors, Middletons Moore & Bevins.

Cummins J:

1. The objects of the *Occupational Health and Safety Act 1985* are set forth in sec 6 of that Act. They are —

"(a) to secure the health, safety and welfare of persons at work;

(b) to protect persons at work against risks to health or safety;

(c) to assist in securing safe and healthy work environments;

(d) to eliminate, at the source, risks to the health, safety and welfare of persons at work;

(e) to provide for the involvement of employees and employers and associations representing employees and employers in the formulation and implementation of health and safety standards."

2. These are serious matters. The provision by employers of a safe workplace and safe systems of work is a serious matter.

3. Under sec 21(1) of the *Occupational Health and Safety Act 1985* it is provided that an employer "shall provide so far as is practicable for employees a working environment that is safe and without risks to health". That is the basal legislative requirement.

4. The section proceeds in subsec (2) that:

"Without limiting the generality of sub-s. (1), an employer contravenes that sub-section if the employer fails —

(a) to provide and maintain plant and systems of work that are so far as is practicable safe and without risks to health;

(b) to make arrangements for ensuring so far as is practicable safety and absence of risks to health in connection with the use, handling, storage and transport of plant and substances;

(c) to maintain so far as is practicable any workplace under the control and management of the employer in a condition that is safe and without risks to health ..."

(proceeding then so far as is relevant to (e))

"(e) to provide such information, instruction, training and supervision to employees as are necessary to enable the employees to perform their work in a manner that is safe and without risks to health."

By subsec (4)(d), an employer is required —

"so far as is practicable ... [to] monitor conditions at any workplace under the control and management of the employer."

Further, under sec 22 of the Act, an employer is required to —

"ensure so far as is practicable that non-employed persons are not exposed to risks to their health or safety arising from the conduct of the undertaking of the employer."

5. The fundamental consideration is prevention. That is the purpose of the legislation. The standard of compliance and of knowledge is objective. In deciding what is practicable as provided in sec 21 and 22, what is looked to is (sec 4):

"(a) the severity of the hazard or risk in question;

(b) the state of knowledge which the employer had or ought to have had about the hazard or risk and any ways of mitigating or removing the hazard or risk;

(c) the availability and suitability of ways to remove or mitigate that hazard or risk; and

(d) the cost of removing or mitigating that hazard or risk."

6. "Hazard" means the potential to cause injury or illness. "Risk" means the likelihood of injury or illness arising from exposure to a hazard.

7. What occurred at Gas Plant One at Longford on 25 September 1998 was no mere accident. To use the term "accident" denotes a lack of understanding of responsibility and a lack of understanding of cause.

8. The then Minister, in the second reading speech to the Bill for the *Occupational Health and Safety (Miscellaneous Amendments) Act 1990*, in Hansard (Assembly), 13 April 1989 at p 758 said this:

> "Too often one hears the response, 'But that was an industrial accident'. This carries a connotation of inevitability, which denies the possibility of prevention. Even worse, it implies that an offence that results in a work-related fatality is not as serious as other criminal offences involving fatalities."

9. I agree.

10. The events of 25 September 1998 were the responsibility of Esso; no one else. Their cause was grievous, foreseeable and avoidable. Their consequence was grievous, tragic and avoidable.

11. During the course of the trial Esso made the following formal admissions:

"1. Operators and supervisors at the Longford plant were not trained with respect to and were not aware of —

   (i) whether or not cold temperature hazards could arise in GP 905, GP 922 or the base of the ROD and associated piping from the loss of lean oil circulation;

   (ii) whether or not cold temperature hazards could develop in GP 905 and 922 or the base of the ROD and associated piping;

   (iii) whether or not dangers associated with failure of cold vessels existed in GP 905, GP 922, the base of the ROD and associated piping.

2. There were no written procedures dealing with any such hazards or dangers that may have existed associated with cold temperatures in GP 905, GP 922, the base of the ROD and associated piping."

12. It was evident from the evidence given by witness after witness before me that the loyal employees, including supervisors, of Esso were entirely unaware of the deadly danger lurking at GP 905 on the Friday morning, 25 September 1998, particularly around 12 noon. They were loyally attending to a leak in GP 922 and evident cold on GP 905 and related areas. Only one man knew the dangers. Mr Vandersteen, a fitter in the maintenance section, saw what was evident to be seen and in evidence said this: "I just said, 'Fuck this, I'm out of here'. We jumped on our bikes and we left the area." The explosion occurred immediately thereafter. Mr Vandersteen was not trained by Esso, but was trained by the Navy. It was the Navy, not Esso, who taught him to be aware of such danger. This failure to train in safety is a most serious dereliction. Tragically, two loyal employees, Mr John Lowery and Mr Peter Wilson, were killed in the rupture which occurred on 25 September. Eight persons were seriously injured. Those deaths and injuries are tragic in themselves. They are also a tragic measure of the hazard and risk in a gas processing facility. Longford is a major hazard installation. The potential for injury is great and obvious.

13. As presently the *Crimes (Industrial Manslaughter) Bill* is to be considered by Parliament, I consider it is inappropriate for me to say anything here about the limited penalties under the *Occupational Health and Safety Act 1985* and the limited scope of the Act, other than this: this tragic case once again demonstrates, if it needs further demonstration, the vital importance of workplace safety.

14. After a four-month trial, Esso was convicted by the jury of 11 counts of breaches of sec 21 and 22 of the Act. Each is an indictable offence: sec 47(2).

15. On Count 1, Esso was convicted of a breach of sec 21(2)(b) of the Act: that the company at Longford between 1 January 1993 and 25 September 1998 failed to provide and maintain so far as was practicable for employees a working environment that was safe — in that it failed to conduct any adequate hazard identification at Gas Plant 1. Esso in fact conducted a hazard identification, namely a Hazop, on Gas Plant 2 and on Gas Plant 3, but failed in that period to conduct one on the oldest part of the plant, Gas Plant 1, and failed to do so even in the face of Exhibit 12B before me, Exxon's Instructions to Affiliates to conduct a retrospective Hazop on plants more than 20 years old. Gas Plant 1 commenced operation in 1969. The identification of hazards in a major hazard installation is obvious and fundamental.

16. On Count 2, Esso was convicted of a breach of sec 21(2)(b) of the Act: that it failed between 1 January 1994 and 25 September 1998 to provide and maintain so far as was practicable for employees a working environment that was safe — in that it failed to conduct any adequate risk assessment of Gas Plant 1. Esso in fact did conduct a risk assessment in December 1994, but that was neither timely for 1998 nor comprehensive, and was self-limiting because it anticipated the Hazop which was budgeted for, but never occurred, on Gas Plant 1.

17. On Count 3, Esso was convicted of a breach of sec 21(2)(a) of the Act: that it failed to provide and maintain plant that was safe at Longford on 25 September 1998 — in particular that vessels in the ROD/ROF area at GP 1 were dangerously cold on the morning of 25 September 1998 and susceptible to failure.

18. On Count 4, Esso was convicted of a breach of sec 21(2)(a) of the Act: that it failed to have any adequate procedures to enable employees to safely respond to loss of lean oil circulation — being written procedures to ensure that employees knew what to do in a situation of uncertainty or crisis.

19. On Count 5, Esso was convicted of a breach of sec 21(2)(e) of the Act: that it failed adequately to train employees to enable them to safely respond to a loss of lean oil circulation, and in particular in relation to the development of cold temperatures in the ROD/ROF area.

20. On Count 6, Esso was convicted of a breach of sec 21(2)(a) of the Act: that it failed to provide and maintain plant and systems of work that operated at safe temperatures — in particular that there were no systems of work which specified critical operating parameters, no written or oral instructions to employees not to exceed such critical operating parameters, and no adequate means of determining the operating temperature of relevant items of plant.

21. On Count 7, Esso was convicted of a breach of sec 21(2)(e) of the Act: that it failed adequately to train employees regarding risks associated with failure of plant at cold temperatures — in particular that it failed to train employees regarding the risks associated with operation of plant below safe temperatures including the dangers associated with, and the potential for, catastrophic failure. I shall return to this count.

22. On Count 8, Esso was convicted of a breach of sec 21(2)(e) of the Act: that it failed to provide adequately trained supervisors capable of safely responding to a loss of lean oil circulation and the plant operating at cold temperatures.

23. On Count 9, Esso was convicted of a breach of sec 21(4)(d) of the Act: that it failed to monitor the conditions — in particular failing to have properly functioning instruments which would longitudinally record what was occurring at the plant to show the history and development of the conditions, a serious breach in a major hazard installation.

24. On Count 10, Esso was convicted of a breach of sec 21(2)(c) of the Act: that it failed to maintain a safe workplace, by failing to have a preventative mechanism which would have precluded a process problem turning into a catastrophe.

25. Finally, on Count 11, Esso was convicted of a breach of sec 22 of the Act: that it failed to ensure the safety of persons other than its employees after the rupture of Gas Plant 905 and the explosions and fires which followed — notably police, ambulance and Country Fire Authority (CFA) personnel.

26. I shall not in this sentence rehearse the detail of the events of 25 September 1998 or their antecedents. They are well known and well documented.

27. During the course of the trial, further documentation appeared — notably Exhibit 61.2, the letter of 30 May 2000 of Esso over the hand of its solicitors to its proposed experts, a letter to which I shall return.

28. I pay compliment to the jury and to the jury system. This jury sat loyally from 5 March to 28 June this year. The jury was constituted of conscientious, responsible citizens. They were, as juries daily demonstrate, also perceptive. This jury, and juries generally, well understand evidence, including

complex evidence when properly presented, as it was here. Further, juries fulfil a fundamental democratic function that it is the community itself which judges serious cases, guided by the law as stated by the trial judge.

29. In presentation of material in the plea on behalf of Esso, learned senior counsel, Mr Titshall, referred to Esso's otherwise very good safety record. Credit should be given where credit is due. Mr Titshall elaborated that the foundation of Esso's safety management system was and is its Operations Integrity Management System, the manuals of which were in evidence. Mr Titshall rightly pointed out Esso's otherwise very good safety record in the petroleum industry. Between 1992 and 1996 Esso employees worked 12 and a half million employee hours without a single lost-time injury, a technical definition involving fatality, permanent disability or time lost from work. He placed before me as Exhibit E1 data of that otherwise very good safety record. I acknowledge that record. He pointed out, rightly, that Esso is involved in heavy industry, including the gas plant concerned, and that that record stands to Esso's credit. He pointed out that since these matters of 25 September 1998, at Longford over 1.7 million work hours have been completed without any lost-time injury. Further, the company has undertaken safety initiatives of numerous sorts which Mr Titshall elaborated: introduction of safety promoters, health and safety representatives, safety leader training courses, accident reporting systems, written expectations, safety awareness tests and other matters and general high standard job training and competency-based assessments.

30. Esso also has received a number of safety awards. It has a five-star rating award from the National Safety Council in 1991, APIA safety awards for 1994, 1995, 1996, 1999 and 2000, being the best safety performance of large companies, and the Fluor Daniel corporate tri-star award for 100,000 hours accident-free in 1999. It also has other awards and commendable safety records including 100,000 mishap-free flying hours of helicopters to platforms. Credit there is properly given to Esso for its otherwise very good safety record.

31. There are three prior convictions relevant here to sentence. At the Magistrates' Court at Geelong on 1 July 1991 Esso was convicted of failing to provide a safe workplace and fined $5,000. The circumstances of that prior conviction were essentially that Esso was the occupier of a service station site at which a vertical LPG tank was being installed at Grovedale near Geelong. It was installed by an unrelated company. A worker — not an employee of Esso but an employee of the installers of the tank — died as a consequence of entering the tank when it contained nitrogen. The installing company and Esso were found by the Coroner to have contributed to the death by failing to provide training in permitting the deceased to enter the confined space. Esso was convicted, as I say, of failing to provide a safe workplace and was sentenced to a fine of $5,000 with costs. There were four charges against Esso. Three were struck out upon Esso pleading guilty to one charge. A particular of that charge was failing to provide training and supervision to its employees, in that case its employee who was the supervisor of the deceased worker.

32. The other two convictions were imposed in the Magistrates' Court at Sale on 25 February 1993 of failing to provide a safe workplace, wherein Esso was sentenced to pay an aggregate fine of $6,000 on those two charges together with costs. That related to a fire on the Tuna platform on 24 April 1989. The Report of 18 November 1991 of the Coroner, Mr G Johnstone, Exhibit 2 on the plea, sets out the circumstances of the events and fire on the Tuna platform in Bass Strait on 24 April 1989. The fire commenced in the main oil line pump engine enclosure and spread to the adjacent service module. The initiating factor was the failure to danger tag a removed valve on the pump while under maintenance. The Coroner found, "The precise reason for the fire results from inadequate danger tagging, work permit procedures and a failure of communication" (p 6 of the Report). The Coroner emphasised in his Report the critical nature of working with highly flammable and explosive materials. He concluded that, especially because of the hazards and restrictions of platform work, "The failure of Esso to introduce an improved permit, tagging lock-out system results in a contribution to

the fire" (p 11). There were six charges. Esso consented to summary jurisdiction, pleaded guilty to two counts and four were struck out. That incident and fire did not involve gas processing. The Coroner was involved because there was a fire. No lives were lost, although there were some injuries.

33. Those prior matters were significantly earlier than the 1998 Longford matters for which Esso now is to be sentenced. However, they are relevant because they demonstrate the importance of workplace safety, the importance of safe systems of work, and the importance of training and supervision.

34. I turn to the proper construction of sec 53(a)(i) of the Act as to the imposition of further penalty.

35. That section operates to provide a discretion to impose an additional penalty on a person convicted of an offence under the Act if a person has previously been convicted of an offence under the Act. Its operation is not limited to a prior conviction for the same offence or even the same category of offence. The power given is discretionary, that is the court may, not must, impose an additional penalty. Section 53 does not increase the maximum penalty for any particular offence the subject of present conviction; rather it empowers the imposition of an additional penalty. The section thus operates globally, is discretionary and is additional. The section does not state criteria for its exercise. Accordingly it falls to be interpreted in accordance with its terms, content and purpose and in accordance with fundamental sentencing principle. The first such fundamental principle is that the offender is not to be punished twice for the one offence. The further penalty is not a second penalty for the prior conviction. Rather, it is a further penalty for the present conviction by reason of the existence of the prior conviction. The further penalty marks the seriousness of present offences in the context of an offender who has previously offended.

36. At common law, sentencing ultimately is an holistic, not segmented, process. As Adam and Crockett JJ said in *R v Williscroft and Ors* (1975) VR 292 at p 300:

> "Ultimately every sentence imposed represents the sentencing judge's instinctive synthesis of all the various aspects involved in the punitive process."

It is not correct to sentence by a sequential, segmented, stepped process. However, given the terms of sec 53 — "in addition to", "further" — it is necessary if sec 53 is invoked to identify the further penalty. The parties are entitled to know it. As a consequence, in the otherwise holistic synthesis, if sec 53 is invoked the element of prior conviction — a narrower concept than character — necessarily must be omitted to ensure there is no double counting. In all other respects the ultimate sentence remains holistic. With every respect, I consider the analysis stated in *DPP v Pacific Dunlop Ltd* (28 June 1994, County Court Victoria) is too restrictive. Under sec 53 there is not a two-step process as there stated.

37. Circumstances justifying the operation of sec 53 are the nature and number of prior convictions, their proximity or remoteness in time to the present offence, their relevance, the character otherwise of the offender, and whether the combination of prior convictions and present conviction demonstrates systemic failure by the offender or a longitudinal, general or flagrant failure to fulfil the lawful obligations of safety in employment.

38. Applying those matters to the present case, on the one hand Esso has an otherwise very good safety record. On the other hand, the prior convictions are clearly relevant because they relate to matters — the provision of a safe workplace, the knowledge of hazard and the importance of training as a preventative means — which are the essence of the present counts, although in different factual circumstances. Accordingly, I consider it is necessary and appropriate to invoke the discretion provided by sec 53. However, given Esso's otherwise very good safety record, I consider that the further penalties imposed should be moderate. I shall allocate the penalties equally to the most relevant present convictions, those that comprehend identification of provision of hazard, Count 1; provision of safe plant, Count 3; and training of supervisors, Count 8.

39. Of general common law principle applicable in sentencing, punishment and general deterrence are of major significance in this case.

40. However, before imposing sentence on Esso it is unfortunately necessary to examine the litigious conduct of Esso in these proceedings. It is necessary both of itself and as an incident of sentencing — remorse and rehabilitation being relevant to that end.

41. Esso was charged with criminal offences, 11 crimes under the *Occupational Health and Safety Act 1985*, and it was fully entitled to defend those charges. The burden of proof at all times was on the prosecution. The prosecution brought the charges; it had the burden of proving them.

42. Esso and its senior officers have expressed remorse for the tragic loss of life and injury which occurred as a consequence of the rupture on 25 September 1998. I have no doubt that that personal remorse is genuine, from and including the Chairman and Managing Director, Mr RC Olsen, down. I acknowledge that genuine remorse.

43. However, personal expressions of remorse need to be translated into reality. In the present case, they have not been. There are three matters which militate against corporate remorse.

44. First, Esso's litigious treatment of its employees. I found that the Esso employees and personnel who were called before me were most impressive. Although I do not like mentioning some individuals rather than all, the young technical operator, Mr Heath Brew, who was injured in the rupture, before me had a quiet dignity and courage. The plant supervisor, Mr Bill Visser, I found particularly decent and impressive. There were numerous other brave, decent and impressive men from Longford who gave evidence before me. One was the first witness called in the trial, Mr Jim Ward, the control room operator. I was most impressed by the integrity of Mr Ward. He was not asked one question in cross-examination before the jury. How unfortunate it was that, on Esso's instructions, its solicitors submitted to the Royal Commission:

> "Mr Ward was in possession of the necessary information to initiate appropriate action to address the loss of lean oil circulation ... Mr Ward's failure in this respect was due to reasons peculiar to himself."

(Submissions on behalf of Esso Australia by its solicitors to the Longford Royal Commission, 26 April 1999, p 83, para 269.)

The truth is there was only one entity responsible for lack of knowledge on that day: Esso. It, and it alone, should have properly trained the operators and supervisors not only in production, which it did, but also in safety. It, and it alone, failed to do so. Mr Ward and the employees did not fail. Esso failed. These tragic events will always live with these decent, impressive and brave men, a number of whom have received bravery awards, including Mr Visser and Mr Ward, but these events occurred through no fault of theirs.

45. The second matter which militates against corporate remorse was the conduct on Esso's instructions of the defence in the trial. While Esso properly is to be given credit for limiting the issues in the trial and making admissions, and I do give Esso full credit, the defence advanced was one of obfuscation — designed not to clarify, but to obscure. Esso sought to make it appear that the identification of hazard, risk and cause was impossibly difficult. To that end, prosecution experts were cross-examined in technical detail to undermine proof which, with its other hand, Esso was promoting to its own experts: the smoking gun, as Mr Richter called it, the letter of Esso's solicitors of 30 May 2000 produced at the eleventh hour, Exhibit 61.2. The convoluted and obscure question asked by the defence of the defence witness, Dr Baybutt, at pp 4464 to 4466 — all 38 lines of it — and the convoluted and obscure scenario posited by Dr Baybutt at pp 4657 to 4659 — all 58 lines of it — are testimony to the defence of obfuscation; and the words were hollow when, as was put in the question at p 4466, line 5, "That's about as simply as I can put it", and by Dr Baybutt in the answer at p 4657, line 27, "I'll try to keep it as simple as I can".

46. The third matter, and the really significant one, is the lamentable failure of Esso to accept its responsibility for these tragic events. Early in the plea learned senior counsel for Esso said this (p 5925, line 27): "Whilst Esso does not accept the accuracy or correctness necessarily of each of the criticisms levelled by the Royal Commission, I will seek to demonstrate that it has taken very positive steps to address each and every one of those criticisms as if it were accepted, and has done so very thoroughly. I will take your Honour to that detail, but first and most importantly, your Honour, again, Esso does wish to once again repeat its most profound regret for the loss of life and injuries that were caused to its employees by the explosions of 25 September 1998. It might be said that that is trite." I asked, "Does Esso accept responsibility for the fatal and injurious events?" Senior counsel replied, "It accepts responsibility in the way that it has thus far." I said, "That simply means what you have just said. You heard what the jury has said and you heard what the Royal Commission said. I'm not talking about the level of hearing. I'm asking the direct question: does Esso accept responsibility for the fatal and injurious events?" Senior counsel replied, "I can't answer that question in any blanket fashion, your Honour", to which I replied, "Very well".

47. Then towards the end of plea, after senior counsel rightly had reviewed Esso's otherwise commendable safety record, I asked the following (p 5970, line 2): "I think that leads to this, Mr Titshall. I have no doubt that the many and commendable facts that you have put before me are accurate and true as to your client, but the corollary of what you put to me is this. In relation to the rupture on 25 September and its antecedents, there is a mountain of evidence; a most distinguished former High Court judge with his co-Commissioner made clear and unequivocal findings, as stated in the Royal Commission Report; and a jury of 12 unanimously found proof beyond reasonable doubt of every charge laid against Esso. If Esso is the good corporate citizen you have stated it to be, why does it not accept responsibility for the fatal and injurious events?" Senior counsel replied, "I can't answer that, your Honour", and I replied, "Very well".

48. Normally in sentencing, a failure to accept responsibility sounds in the matter of specific deterrence. This is not so here, because Esso has demonstrated by its subsequent operational reforms that it has acted responsibly to remedy past deficiency. But its lack of acceptance of responsibility has a relevance in my not accepting that Esso's expression of remorse is practical and operational. Esso's failure still to accept responsibility for these tragic events is a serious deficiency.

49. Esso has been convicted on 11 counts. Each count is a separate offence. There is an interface — not a duplication — of some counts. Accordingly, I impose penalty on a classification basis. Counts 1 and 2 relate to each other: Count 1, the identification of hazard in a most hazardous workplace, and Count 2, the assessment of risk consequent upon the identification of hazard. On the facts of this case, Counts 3, 6, 7 and 8 relate to the matter generally of excessive cold. Counts 4 and 5 relate to the loss of lean oil. Count 9 relates to monitoring. Count 10 is a consequent count. Count 11 relates to non-employees, in particular police, ambulance and CFA personnel. Specifically to ensure there is no double punishment and no double counting, I moderate the sentences on the individual counts each to the other in those categories, as well as generally, complying with the principle of totality.

50. Given the purpose, nature and provisions of this legislation and the facts of this case, all the offences are serious. Two warrant the maximum penalty. Count 1 warrants the maximum penalty: the failure by Esso over time to conduct hazard identification in a most hazardous workplace. Count 7 warrants the maximum penalty: the failure by Esso to train its employees about the risks they were subject to. Those employees and their loved ones had every right to be properly trained about the risks Esso sent them to face. Esso totally failed in that most fundamental of matters. I impose the maximum penalty on those two counts. Some counts warrant the imposition of a very substantial penalty, although less than the maximum. Count 3 is such: Esso's failure in the fundamental requirement of providing and maintaining safe plant. Count 9 is another: Esso's failure in the essential

requirement of monitoring a hazardous workplace. Count 11 is another: Esso's failure in ensuring the safety of outside persons. All the training counts are serious, because proper training is at the front line of prevention and prevention is the essence of workplace safety. However, there is some interface of the various training counts. I have moderated the penalty on those counts to reflect that interface. That does not connote lack of seriousness. Proper training is vital.

51. Bearing in mind the above principles, criteria and considerations, I impose sentence as follows:

52. On Count 1, failure to conduct any hazard identification at Gas Plant 1, I impose the maximum penalty, a fine of $250,000. Pursuant to sec 53(a)(i) of the *Occupational Health and Safety Act 1985*, on Count 1 I impose a further penalty of a fine of $50,000, making a total penalty on Count 1 of a fine of $300,000.

53. On Count 2, failure to conduct any adequate periodic risk assessment at Gas Plant 1, I impose a fine of $150,000.

54. On Count 3, failure to provide and maintain plant that was safe, I impose a fine of $200,000. Pursuant to sec 53(a)(i), on Count 3 I impose a further penalty of a fine of $50,000, making a total penalty on Count 3 of a fine of $250,000.

55. On Count 4, failure to have adequate response procedures, I impose a fine of $100,000.

56. On Count 5, failure to adequately train employees to respond safely to loss of lean oil circulation, I impose a fine of $100,000.

57. On Count 6, failure to provide and maintain plant that operated at safe temperatures, I impose a fine of $100,000.

58. On Count 7, failure to adequately train employees about risks, I impose the maximum penalty, a fine of $250,000. Pursuant to the provisions of sec 53(a)(i), on Count 7 I impose a further penalty of $50,000, making a total penalty on Count 7 of a fine of $300,000.

59. On Count 8, failure to provide adequately trained supervisors, I impose a fine of $150,000. Pursuant to the provisions of sec 53(a)(i), on Count 8 I impose a further penalty of $50,000, making a total penalty on Count 8 of $200,000.

60. On Count 9, failure to monitor conditions, I impose a fine of $200,000.

61. On Count 10, failure to prevent the rupture and explosions by a preventative mechanism, I impose a fine of $100,000.

62. On Count 11, failure to ensure the safety of non-employees, particularly police, ambulance and Country Fire Authority personnel, I impose a fine of $200,000.

63. In total, for the 11 offences I fine Esso Australia Pty Ltd the sum of $2,000,000.

64. I grant a stay of 28 days for payment of the fines imposed.

65. *Sine die.*

INDEX

Accident prevention
charges against Esso ... 11

Alarms, ignoring ... 45-46

Anonymous reporting system
safety improvements since explosion ... 46-47

Australian Standard AS/NZS 4360 ... 15-16; 55

Australian Vinyls safety case ... 43; 47

Blameworthiness ... 6; 9; 32

Brittle fractures — see **Cold metal embrittlement**

Cause of disaster
Esso's defence ... 28-29
new findings ... 55

Charges against Esso
choice of court ... 11
eleven counts ... 13; 56

Cold metal embrittlement
inadequate hazard identification ... 13-15
inadequate risk assessment ... 16
inadequate training of employees/supervisors ... 18-19
questioning the causal connection ... 28-29

Control measures
performance indicators ... 42-43

Control room operators
not blamed at trial ... 7; 57

Corporate behaviour ... 9

Corporate manslaughter ... 49-54; 58

Corporate remorse ... 7; 36-37; 57

Country Fire Authority
personnel placed at risk ... 21-22

***Crimes (Workplace Deaths and Serious Injuries) Bill 2001* (Vic)** ... 50-54

Criminality ... 7; 27; 50-51

Critical events ... 43-44

Critical operating procedures and alarms ... 45-46

Culpability — see **Blameworthiness**

Defence counsel ... 8

Defence for Esso
blame of operator ... 32
causal connection ... 28-29
culpability ... 27
effectiveness of strategy ... 56
position on cause of rupture ... 29-30

Deterrence ... 34-35

Embrittlement — see **Cold metal embrittlement**
Emergency workers at risk 21-22
Engineers
absence at site 19-20
safety improvements since explosion 46
Esso
corporate remorse 36-37
criminality of offences 7
culpability 57
defence 27-32
lessons learned from Longford 7; 57-58
lost-time injuries 41-45
multiple charges 6
prior convictions 37-39
reputation 5
risk of corporate manslaughter conviction 7; 58
Royal Commission and trial compared 7-9
safety awards 44-45
safety improvements since explosion 45-46
safety management system 44
Exxon 5; 52

Fatalities
corporate manslaughter 49-54
Fines — see **Penalties**
Foreseeability
providing a safe workplace 22-23
risk, sentencing 34

Gas processing industry 5
Gas supply loss 6
Gross negligence 50; 52-53

Hazard
confusion with risk 7; 15; 55
Hazard and operability study — see **HAZOP**
Hazard identification
inadequacy 13-15
safety improvements since explosion 46
Hazard management process
charges against Esso 11
HAZOP (hazard and operability study)
Esso's failure to conduct 44; 52-53
Esso's questioning foreseeability 30-32
hazard identification 14
risk assessment 17; 24
safety improvements since explosion 46

Heat exchanger — see also **Cold metal embrittlement**
failure to protect from thermal shock 20-21

Incident reporting — see **Reporting procedures**
Industrial manslaughter 49-54; 58
Injury management strategies 42

Legislation — see ***Occupational Health and Safety Act 1985*** **(Vic)**
Longford disaster
new findings 7; 55
Lost-time injuries 41-45

Magistrates' Court 23
Major hazard facilities 42
Major hazard guidance note 14; 42; 44
Mens rea **(criminal state of mind)** 27; 50-51
Metal fractures — see **Cold metal embrittlement**
Monitoring of plant, failure 19-20
Multiple offences 23-25

Near misses 46
Negligence 49-54
Nuclear industry 5

Occupational Health and Safety Act 1985 **(Vic)**
ensuring safety of non-employees 21
Esso's violation 8
OHS offences
criminality 27; 50-51; 56
multiple 23-25
Oil and gas industry 5
Operations Integrity Management System ..44
Operator error 7
Operator log books 47
Organisational change 5
Organisational culture 47

Penalties
corporate manslaughter 49-54
Esso conviction 34
multiple offences 24
purpose of punishment 34-36
Shell Australia 6

Performance indicators
control measures 42-43

Performance monitoring 42-44

Periodic risk assessment 16-17

Petrochemical industry 14

Piper Alpha..........42

Plant
failure to ensure operation at safe temperature 18
failure to maintain safety 17

Plea 37

Practicability
providing a safe workplace..........22-23

"Practicably preventable" 7; 56

Prior convictions 37-39; 57

Procedures, absence
warm oil circulation 17-18

Process upsets
new focus by Esso..........45

Prosecution
incident sequence 12
multiple counts against Esso 23-25; 56

Public interest..........11

Punishment — see **Sentencing**

Recording practices..........42-44

Rehabilitation..........35-36

Remote monitoring
safety improvements since explosion 46

Reporting procedures 43-44
anonymous reporting system 46-47

Reputation of Esso 5

Risk
confusion with hazard 7; 15; 55
defined by AS/NZS 4360 ....16; 55

Risk analysis 15-16

Risk assessment
inadequate 15-17

Risk control..........15

Risk identification 15

Risk management 15

Risk treatment 16

Royal Commission
comparison with trial 7-9
incident sequence 12

Safe workplace 22-23

Safety awards..........41; 44-45

Safety management systems
Esso's safety record..........41
Operations Integrity Management System 44
Shell Australia..........6

Safety-critical incidents 47

Sentencing — see also **Penalties**
corporate remorse 36-37
Judge's statement 33
multiple charges 39
plea..........37
prior convictions..........37-39; 57
purposes of punishment 34-36

Shell Australia 6

State Government
corporate manslaughter legislation 53

Supervisors
additional position since explosion 46
inadequate training for risks of
abnormally cold plant..........19

Supreme Court of Victoria
choice of court for Longford trial..........55
multiple counts against Esso 23-25

Table top drills 45; 46

Thermal shock
failure to protect heat exchanger 20-21

Training
failure associated with cold plant..........18-19
failure to respond to loss of warm oil 18
safety improvements since explosion..........45
supervisors and risks of abnormally cold plant......19

Trial
comparing with Royal Commission..........7-9
Esso's multiple counts 23; 56
multiple charges 6
multiple offences..........23-25

Victorian WorkCover Authority 6

Warm oil system
failure to train employees..........18
inadequate procedures..........17-18
shutdown of pumps..........12

WorkCover 6

NOTES FOR CONTRIBUTORS

Contributions are always welcome. News, letters and articles should be sent to:

Managing Editor
The Journal of Occupational Health and Safety
GPO Box 4072
Sydney, NSW 2001

Contributions can be discussed with the Managing Editor at CCH Australia Limited (contact ☎ (02) 9857 1845).

New Zealand contributors should note that material for the journal may be sent to:

CCH New Zealand Limited
24 The Warehouse Way
Northcote 1309, Auckland 10

marked for forwarding to CCH Australia Limited.

Articles with original material are accepted for consideration with the understanding that, except for abstracts, no part of the data has been published, or will be submitted for publication elsewhere, before appearing in this journal. Authors are required to assign copyright to CCH Australia Limited when their article is accepted for publication.

INSTRUCTIONS TO CONTRIBUTORS

Articles should be about 3,000 words (longer articles will be considered) and carry an abstract of not more than 150 words, stating the key points of the material. Please supply five keywords and brief details of the author's professional qualifications, current position and employer.

OHS in brief articles are short reports of less than 1,500 words.

Letters should not exceed 300 words.

Conferences/seminars/courses. Details supplied for publication should include date, time, venue and contact persons.

PREPARATION OF MANUSCRIPTS

Manuscripts should be written in concise language and should be typed double spaced, using only one side of the paper. Number pages consecutively and leave wide margins. A separate title page should contain the title, the author's full name and qualifications, and details relevant to correspondence. If possible, include a word count.

Authors should submit one original and two copies of each manuscript. It is good practice for the author to retain his/her own copy of the manuscript. Articles will be submitted for expert peer review (with authors and referees remaining anonymous).

On acceptance of the article, authors will be asked to provide the manuscript on disk. The preferred form is Word for Windows 7.0 or an ASCII (plain text) file on a PC DOS or MS DOS formatted disk.

Illustrations, tables and graphs. All illustrations (line drawings, graphs, charts, diagrams, photographs) should be suitable for easy reproduction. Photographs should be submitted as glossy black and white prints. Type legends in double spacing on separate pages. To assist anonymous editorial review, do not indicate author names on illustrations.

Statistical data should be expressed in SI units.

References should be cited in the text by superior numbers and a full list of references given at the end of the manuscript in numerical sequence (based on the Vancouver system). References to books should include authors' surnames and initials, full title, place of publication, full name of publisher and date of publication. References to journal articles should include authors' surnames and initials, full title of article, full title of journal, date of publication, volume number, issue number and page span.

The accuracy of references is the author's responsibility. Check each reference in the manuscript and again in the proofs.

Authors are responsible for ensuring that their words do not infringe copyright.

House style. Manuscripts will be edited following CCH house style. These changes will be at the discretion of the publisher.

Alterations to proofs. The only acceptable changes at this stage of production are to correct misprints or errors of fact. Major alterations of wording or the addition of new material cannot be accepted at proof stage. Any queries written in the margins of proofs need to be answered. Page proofs must be returned promptly.

Detailed *Guidelines for Authors* may be obtained from the Managing Editor.